Achieving Water Security

Praise for the book

'Water security matters, and it matters most to the most vulnerable individuals, households, communities and nations. Ethiopia is home to many millions of people who are anything but water-secure. This publication shines a light on such people, and it provides numerous inter-disciplinary insights into the causes of water insecurity and into some of the measures which are needed to relieve that insecurity. There are too few research programmes of this type which focus on situations of real need in the world, and which grapple with the difficult issues associated with the relief of poverty and vulnerability. I warmly welcome this contribution to the literature, and hope that this work will inform the practices and policies of government and development partners in Ethiopia.'

Richard C. Carter, Visiting Professor, Cranfield University, UK

'Consolidating these findings from the RiPPLE programme into a single volume to address water security in Ethiopia has generated an essential primer and benchmark in realizing domestic water supply and sanitation in Africa's second most populous country. Being informed by five years of research and learning, this is a valuable insight into a critical dimension of Ethiopia's future security.'

Melvin Woodhouse, Associate Director, LTS International, UK

Achieving Water Security

Lessons from research in water supply, sanitation and hygiene in Ethiopia

Edited by
Roger Calow, Eva Ludi
and Josephine Tucker

PRACTICAL ACTION
Publishing

Practical Action Publishing Ltd
The Schumacher Centre
Bourton on Dunsmore, Rugby,
Warwickshire CV23 9QZ, UK
www.practicalactionpublishing.org

Copyright © Overseas Development Institute, 2013

ISBN 978 1 85339 7639 Hardback
ISBN 978 1 85339 7646 Paperback
ISBN 978 1 78044 763 6 Library Ebook
ISBN 978 1 78044 764 3 Ebook

All rights reserved. No part of this publication may be reprinted or reproduced or utilized in any form or by any electronic, mechanical, or other means, now known or hereafter invented, including photocopying and recording, or in any information storage or retrieval system, without the written permission of the publishers.

A catalogue record for this book is available from the British Library.

Roger Calow, Eva Ludi and Josephine Tucker, eds (2013)
Achieving Water Security: Lessons from Research in Water Supply, Sanitation and Hygiene in Ethiopia, Practical Action Publishing, Rugby.

The contributors have asserted their rights under the Copyright Designs and Patents Act 1988 to be identified as authors of their respective contributions.

Since 1974, Practical Action Publishing (formerly Intermediate Technology Publications and ITDG Publishing) has published and disseminated books and information in support of international development work throughout the world. Practical Action Publishing is a trading name of Practical Action Publishing Ltd (Company Reg. No. 1159018), the wholly owned publishing company of Practical Action. Practical Action Publishing trades only in support of its parent charity objectives and any profits are covenanted back to Practical Action (Charity Reg. No. 247257, Group VAT Registration No. 880 9924 76).

This material has been funded by UK aid from the UK Government, however the views expressed do not necessarily reflect the UK Government's official policies.

The opinions expressed in this book are those of the authors, and do not necessarily reflect the views of their institutions.

Cover photo: A girl on her way to collect water from a pond
Photo credit: Iva Zimova / Panos
Indexed by Liz Fawcett, Harrogate, North Yorkshire
Typeset by Bookcraft Ltd, Stroud, Gloucestershire
Printed in the UK by Short Run Press

Contents

Photos and figures vii
Tables viii
Boxes ix
Foreword x
Acknowledgements xii
Acronyms xiii
Glossary xvii

   Introduction 1
   Roger Calow, Zemede Abebe and Alan Nicol

1 Ethiopia's water resources, policies, and institutions 25
   Eva Ludi, Bethel Terefe, Roger Calow and Gulilat Birhane

2 WASH sector monitoring 49
   John Butterworth, Katharina Welle, Kristof Bostoen and Florian Schaefer

3 Innovative approaches for extending access to water services: the potential of multiple-use water services and self-supply 69
   Marieke Adank, John Butterworth, Sally Sutton and Zemede Abebe

4 Sanitation and hygiene promotion in rural communities: the Health Extension Programme 89
   Peter Newborne and Anu Liisanantti

5 Sustainability of water services in Ethiopia 107
   Nathaniel Mason, Alan MacDonald, Sobona Mtisi, Israel Deneke Haylamicheal and Habtamu Abebe

6 Water for livelihood resilience, food security, and poverty reduction 127
   Josephine Tucker, Zelalem Lema and Samson Eshetu Lemma

7 Responding to climate variability and change: implications for planned adaptation 147
   Lindsey Jones, Lorraine Coulter, Million Getnet Gebreyes, Beneberu Shimelis Feleke, Naomi Oates, Leulseged Yirgu Gebreamlak and Josephine Tucker

8   Policy and practice influence through research: critical
    reflections on RiPPLE's approach   173
    Josephine Tucker, Ewen Le Borgne and Marialivia Iotti

    Index   195

Photos and figures

Photos

Pumps drawing irrigation water from the Awash River	28
Collecting water for domestic use from the irrigation system in Ifa Daba	77
Contrasting experiences of irrigation (in Box 6.1)	139

Figures

1.1	Schematic map of Ethiopia's river basins	27
1.2	Proportion of national populations without access to improved drinking water	29
2.1	Rural water supply coverage	51
3.1	Pathways towards multiple-use water services	75
3.2	Benefit/cost and incremental benefit/cost ratios	76
3.3	Comparison of water quality in different source types (wet season)	77
4.1	A typical 'sanitation ladder' in rural SNNPR	94
5.1	Conceptual framework for sustainability of water services	109
5.2	The sustainability of water points related to aquifer productivity for three *woredas* in Benishangul-Gumuz	118
5.3	Estimated groundwater storage in Africa	120
6.1	Causal pathways linking lack of water with food insecurity	128
6.2	Incidence and depth of poverty among households with and without improved water supply in East Hararghe	131
6.3	Seasonal calendars of water access	134
7.1	Rainfall distribution in Ethiopia	150
7.2	Coarse-resolution GCM outputs showing projected changes in temperature at a sub-national scale	152
7.3	Water use for human consumption as a percentage of minimum human needs in the highland Wheat Barley Potato (WBP) livelihood zone	155
7.4	Water use by livestock in the lowland Shinile Agropastoral (SAP) livelihood zone	156
8.1	Linear technology transfer model versus multi-stakeholder learning process	178
8.2	RiPPLE's research studies and influencing areas	179

Tables

1.1	Water-related targets in the Growth and Transformation Plan	30
1.2	Water supply and sanitation coverage and investment figures	35
1.3	Non-functionality of rural water schemes in 10 regions	36
2.1	National and globally-reported rural and urban water supply coverage figures for Ethiopia, 2010	51
2.2	Compilation of national and internationally-reported data points for rural water supply coverage	53
4.1	Sanitation and hygiene – elements of the Health Extension Programme	92
6.1	Experience of water-related disease among different age groups	131
6.2	Experience of diarrhoeal disease among individuals with differing access to improved water and handwashing practices	132
8.1	RiPPLE strategies for policy influence: tackling common challenges	175
8.2	Major RiPPLE influencing strategies beyond the LPA	181

Boxes

1.1	Ethiopian Water Resources Management Policy (1999)	31
1.2	Financing modalities for WASH	33
1.3	Financing WASH services in East Hararghe: findings from RiPPLE research	34
1.4	Institutional responsibilities under the WASH MoU	38
1.5	Key principles in WASH implementation	43
2.1	Coverage as defined by the Ministry of Water and Energy	52
2.2	Data collected in the National WASH Inventory	57
2.3	Rural water supply in the National WASH Inventory	58
2.4	Encouraging local use of data	60
3.1	Recent milestones and policy developments for self-supply	79
3.2	Self-supply and multiple uses in Mirab Abaya	84
5.1	What is a water service?	109
5.2	Unheard voices	113
5.3	Implications of climate change for water services sustainability	119
6.1	Contrasting experiences of irrigation	139
7.1	Summarized projections of future global climate trends related to freshwater resources	148
7.2	Farmers' perceptions of change	151
7.3	Summary of GCM ensemble projections for Ethiopia	151
7.4	National Adaptation Programme of Action (NAPA)	158
8.1	The struggle for information	174
8.2	Action research	176
8.3	Learning and Practice Alliances	177
8.4	Enhancing scheme sustainability in two *woredas*	182
8.5	Expanding water services for multiple uses	182
8.6	Improving life for farmers: equitable water supply and better prices for vegetables	183
8.7	Influencing the Universal Access Plan and National WASH Inventory	184
8.8	Engaging on food security, climate change, and disaster risk management strategies	184
8.9	Improving scheme sustainability by building WASHCO capacity	185
8.10	Practical planning skills for *woreda* water officers	187
8.11	Embedding research approaches in national universities	188

Foreword

Since 2006, the DFID-funded RiPPLE programme has provided invaluable support to the water and sanitation sector in Ethiopia. Drawing on the capabilities of local staff and working closely with the Government of Ethiopia, RiPPLE has made a significant contribution to capacity building, the generation of new knowledge, and policy and practice around the delivery of water and sanitation services in rural areas. This is a major achievement, and some of the key highlights and findings are captured in this book.

The challenge in meeting the water and sanitation needs of a growing population is a huge one, and climate change will make the challenge greater. But significant progress is being made, not least with the acceleration in water supply coverage under the Government of Ethiopia's Growth and Transformation Plan.

As RiPPLE enters its next phase as an independent organization, managed fully by Ethiopian staff, it can play a key role in supporting this effort: conducting high-quality research and providing evidence that helps extend and sustain access to water and sanitation services, and builds a water-secure future for all Ethiopians.

H.E. Alemayehu Tegenu
Minister, Ministry of Water and Energy, Government of Ethiopia
Addis Ababa, September 2012

When RiPPLE began work in 2006, it could not have been anticipated that an externally funded research programme would become an independent organization with a unique voice in Ethiopia's water and sanitation sector. Yet this is the transformation that has occurred, and I am honoured to be leading RiPPLE as it charts new waters with a dedicated team of Ethiopian staff.

In November 2011, RiPPLE was registered as an independent NGO under the Ministry of Justice, and has since grown in size and strength. Unlike many of the authors of this book, I am a relative newcomer to RiPPLE, taking over the reins as Director in April 2012 from my esteemed predecessor, Zemede Abebe. Looking ahead, there is clearly much to do, building on what has been achieved and documented in these pages, but enhancing further our position as a national resource centre, and as a provider of research-based evidence, policy advice and capacity building. In so doing, we will be engaging with the up and coming issues around climate-compatible development, disaster risk reduction and water resources management, as well as with the ongoing challenge of extending – and sustaining – access to water and sanitation services. These are issues that can only grow in importance as Ethiopia seeks to safeguard the progress it has already made, while striving to achieve middle-income status.

This book demonstrates what we can do, and I commend the authors for their achievement in summarizing such a rich body of research over five years. As for the next five years, RiPPLE will redouble its efforts to provide robust evidence and advice for all sector stakeholders, with the ultimate aim of ensuring all Ethiopians achieve water security.

Deres Abdulkadir
Director, RiPPLE
Addis Ababa, September 2012

Acknowledgements

This is a book that attempts to capture at least some of the rich insights, findings and mutual learning to emerge from the RiPPLE programme over five years. We're delighted to see final publication after so much hard work from the authors and many contributors.

A full list of people to thank for their work on the book, and their contribution and support for the RiPPLE programme, would run to a dozen pages or more. Instead, we'd simply like to thank all of those who helped make RiPPLE a success, and such a happy programme to work on, from staff in the Ministry of Water and Energy, to the countless individuals in the regions who gave their time freely, and enthusiastically, to the research. And of course to RiPPLE partners and staff, past and present, for their commitment and professionalism throughout. RiPPLE partners deserve a special mention: the International Water and Sanitation Centre (IRC), Delft, The Netherlands; the Hararghe Catholic Secretariat (HCS), Dire Dawa, Ethiopia; WaterAid Ethiopia, Addis Ababa, Ethiopia; and the College of Development Studies (CDS) at Addis Ababa University, Ethiopia.

RiPPLE is now entering a new and exciting phase as an independent organization, and all of those involved can be rightly proud of this achievement. RiPPLE is continuing to play an active and unique role in the sector – as a resource centre, and provider of research-based evidence for policy-makers, practitioners and other stakeholders. We wish it every success.

Roger Calow, Eva Ludi and Josephine Tucker

Acronyms

AfDB   African Development Bank
AMCOW   African Ministers' Council on Water
ASL   Above sea level
BoFED   Bureau of Finance and Economic Development
BoH   Bureau of Health
BoWE   Bureau of Water and Energy
BoWR   Bureau of Water Resources
CAPEX   Capital expenditure
CCRDA   Consortium of Christian Relief and Development Association
CDF   Community Development Fund
CDKN   Climate and Development Knowledge Network
CHP   Community health promoter
CHW   Community health worker
CLTS   Community-led Total Sanitation
CLTSH   Community-led Total Sanitation and Hygiene
CMP   Community-managed project
COWASH   Community-led Water, Sanitation and Hygiene
CRGE   Climate-Resilient Green Economy
CRS   Catholic Relief Services
CSA   Central Statistical Agency
CSO   Civil society organization
CWA   Consolidated Wash Account
DAG   Development Assistance Group
DALY   Disability-adjusted life year
DFID   Department for International Development (UK)
DHS   Demographic and Health Services
DRA   Demand-responsive approach
DRC   Democratic Republic of Congo
DRR   Disaster risk reduction
ECWP   Ethiopia Country Water Partnership
ENSO   El Niño / La Niña–Southern Oscillation
EPA   Environmental Protection Authority
ESRDF   Ethiopia Social Relief, Rehabilitation and Development Fund
ET   Evapo-transpiration
ETB   Ethiopian birr
EU   European Union
FAO   Food and Agriculture Organization

FDRE   Federal Democratic Republic of Ethiopia
FEWS-NET   United States Aid Agency Famine Early Warning Systems Network
FLoWS   Forum for Learning on Water and Sanitation
GCM   Global Circulation Model
GDP   Gross domestic product
GHG   Greenhouse gas
GLoWS   Guided Learning on Water and Sanitation
GoE   Government of Ethiopia
GPS   Global Positioning System
GTP   Growth and Transformation Plan
HEA   Household Economy Analysis
HEP   Health Extension Programme
HEW   Health extension worker
HSDP   Health Sector Development Programme
IDS   Institute of Development Studies
IEC   Information, education, and communication
IFAD   International Fund for Agricultural Development
IKM   Information and knowledge management
INGO   International non-governmental organization
IPCC   Intergovernmental Panel on Climate Change
IRC   International Water and Sanitation Centre
ITCZ   Inter-Tropical Convergence Zone
KM4Dev   Knowledge Management for Development
KWT   Kebele WASH Team
IMR   Infant mortality rate
IWRM   Integrated Water Resources Management
JICA   Japan International Cooperation Agency
JMP   Joint Monitoring Programme
JTR   Joint technical review
KAP   Knowledge, attitude, and practice
LPA   Learning and Practice Alliance
LPCD   Litres per capita per day
MDG   Millennium Development Goal
M&E   Monitoring and evaluation
MOARD   Ministry of Agriculture and Rural Development
MoFED   Ministry of Finance and Economic Development
MoH   Ministry of Health
MoU   Memorandum of Understanding
MoWE   Ministry of Water and Energy
MoWR   Ministry of Water Resources
MSF   Multi-stakeholder forum
MUS   Multiple-use water services
MW   Megawatt
NAPA   National Adaptation Programme of Action

NGO Non-governmental organization
NWI National WASH Inventory
ODA Overseas development assistance
O&M Operation and maintenance
ODI Overseas Development Institute
OECD Organisation for Economic Co-operation and Development
OMSU Operation and Maintenance Support Unit
PANE Poverty Action Network Ethiopia
PCDP Pastoral Community Development Project
PSNP Productive Safety Net Programme
PTA Parent–teacher association
RCCSGA Resource Center for Civil Society Groups Association
REDD Reducing emissions from deforestation and forest degradation
RiPPLE Research-inspired Policy and Practice Learning in Ethiopia and the Nile Region
RWSEP Rural Water Supply and Environmental Programme
RWSN Rural Water Supply Network
SAP Shinile Agropastoral
S&H Sanitation and hygiene
SMC Sorghum Maize Chat
SNNPR Southern Nations, Nationalities, and People's Region
SNV Netherlands Development Organisation
SOAS School of Oriental and African Studies
SSA Sub-Saharan Africa
SWAp Sector-wide approach
TB Tuberculosis
TTC Thermotolerant coliform
TVETC Technical and vocational education and training college
UAP Universal Access Plan
U5MR Under-five mortality rate
UNDP United Nations Development Programme
UNFCCC United Nations Framework Convention on Climate Change
UNICEF United Nations Children's Fund
USAID United States Agency for International Development
VHW Village health worker
VLOM Village-level operation and maintenance
WASH Water, sanitation and hygiene
WASHCOs Water, sanitation and hygiene committee
WBP Wheat Barley Potato
WELS Water Economy for Livelihoods
WIF WASH Implementation Framework
WFP World Food Programme
WHO World Health Organization
WME Water, mining and energy
WMS Welfare Monitoring Survey

WSF-CCRDA Water Sanitation Forum-Consortium of Christian Relief and Development Associations
WSP Water and Sanitation Programme
WUA Water user association
WWF World Wide Fund for Nature
WWRDO Woreda Water Resource Development Office
ZWRDO Zonal Water Resource Development Office

Glossary

Adaptation. Adjustment in natural or human systems to a new or changing environment; adaptation can be anticipatory or reactive, private or public, autonomous or planned

Adaptive capacity. The ability of a system (e.g. a community or household) to anticipate, deal with and respond to change

Bega. Dry season

Belg. Short rainy season

Chat. Tree reaching 25 m when grown naturally, but generally kept to 1.5–4 m, and cultivated as a cash crop; has hallucinogenic properties

Climate change. A statistically significant change in either the mean state of the climate or in its variability, persisting for an extended period (decades or longer)

Climate model. A quantitative approach to representing the interactions of the atmosphere, oceans, land surface and ice (see also Global Circulation Models)

Climate proofing. Ensuring that current and future development policies, investments or infrastructure are resilient to climate variability and change, reducing climate-related risks to acceptable levels

Climate risk. Likelihood of a natural or human system suffering harm or loss due to climate variability or change

Climate variability. The departure of climate from long-term average values, or changing characteristics of extremes, e.g. extended rainfall deficits that cause droughts, or greater than average rainfall over a season

Community Health Promoters (CHPs). Trained volunteers who disseminate health messages, share knowledge with their local communities and encourage good practices

Community management. An approach to service provision in which communities take responsibility for operating and maintaining their own water supply systems

Cost recovery. Recovery of some proportion of investment and/or operation and maintenance costs from service users

Cost sharing. Sharing of costs between different stakeholders involved in service provision, e.g. between communities and government agencies

Coverage. Level of access to a minimum standard of service, usually defined by government

Decentralization. The transfer of tasks, responsibilities and resources to lower level authorities

Declaration of Alma-Ata. Declaration adopted at the International Conference on Primary Health Care (1978), expressing the need for urgent action by all governments, health and development workers, and the international community, to protect and promote the health of all people

Domestic water. Water used by households for drinking, washing and cooking

Ecosystem services. Benefits people obtain from ecosystems. Includes *provisioning* services (e.g. production of food and water); *regulating* services (e.g. flood control); *supporting* services (e.g. nutrient cycling) and *cultural* services (e.g. recreational, spiritual)

Extension workers (health/agriculture). Salaried government staff who provide training and advice to local communities

Food security. When all people, at all times, have access to sufficient, safe, and nutritious food to maintain a healthy and active life

Forum for Learning on Water and Sanitation (FLoWS). Federal-level Learning and Practice Alliance (LPA) established by RiPPLE and hosted by the Ministry of Water and Energy

Functionality (of water systems and services). A measure of whether systems and services are 'fit for purpose' and functioning as intended; typically used to distinguish between systems that work and provide services, and systems that don't because they have fallen into disrepair

Global Circulation Models (GCMs). Global climate models used to project future climates using various scenarios to see how the climate will evolve under certain parameters

Green economy. An economy with significantly reduced environmental risks and ecological scarcities, resulting in improved human well-being and social equity

Green energy. Sources of energy that produce fewer greenhouse gas emissions (or other pollutants), therefore mitigating anthropogenic climate change (or other negative impacts on the environment). Typically from renewable energy sources such as solar power, hydro-power or bio-fuels

Green Revolution. A series of research, development, and technology transfer initiatives, occurring between the 1940s and the late 1970s, that increased agriculture production, particularly in India, based on high-yielding crop varieties and intensive use of inputs

Guided Learning on Water and Sanitation (GLoWS). A practical course developed and piloted by RiPPLE to train local water technicians and health staff in the basics of water supply and sanitation, combining classroom-based teaching with on-the-job training

Hardware. Technology and built infrastructure

Household water economy. The sum of the ways in which a household accesses and uses water to support its livelihood(s)

Improved water supply/source. A source that is likely to be protected from outside contamination, particularly from faecal matter. The WHO/

UNICEF Joint Monitoring Programme (JMP) includes within this category piped water, public taps, boreholes, protected dug wells, protected springs and rainwater

Integrated Water Resources Management (IWRM). A process which promotes the coordinated development and management of water, land and related resources in order to maximize the resultant economic and social welfare in an equitable manner without compromising the sustainability of vital ecosystems

Kebele. Lowest administrative unit in Ethiopia, equivalent to a community and composed of several villages (*Got*)

Kiremt. Main rainy season

Learning and Practice Alliance (LPA). Stakeholder platforms established at different levels (national, regional and district) whose members identify knowledge gaps, and prioritize and participate in research to address them

Mal-adaptation. Changes to a system or human actions that inadvertently increase vulnerability to climate-related hazards; these may be beneficial in the short term but erode adaptive capacity in the longer term

Meher. Crops grown during the main rainy season

Millennium Development Goals (MDGs). A set of eight international development goals that UN member states and international organizations agreed to achieve by 2015

Multiple Use Services (MUS). Water supply systems that incorporate both domestic and productive uses of water in their design and delivery. Multiple services can be provided from a single source or from different sources

Paris Declaration. Declaration on aid effectiveness (2005) attempting to change the way donor and developing countries 'do business' together, based on principles of partnership

Potable water. Water that is safe for humans to drink

Productive water. Water used for economic activities, including livestock watering, irrigation, brick-making, and brewing

Quintal. Unit of measurement, equivalent to 100 kg

Regional Bureau of Water and Energy. Regional-level offices under the authority of the Ministry of Water and Energy responsible for overseeing water resource development, management and service delivery across regional zones and *woredas* (through *woreda* water offices). Sometimes referred to as 'water bureau'

Resilience. The ability of a system and its component parts to anticipate, absorb, accommodate, or recover from the effects of a shock or stress in a timely and efficient manner

River Basin Organizations (RBOs). Institutions set up to coordinate water resources planning and management at basin scale

Robust decision-making. Those decisions made with consideration of uncertainty, such as climate uncertainty. A robust decision will deliver desired benefits under a range of possible scenarios but will not necessarily be the optimal decision for any one single (e.g. climate) scenario

Self-supply (facilitated). Approach to service provision in which the initiative and investment to build or improve water or sanitation sources comes from individual households, usually with some support from external agents

Software. Knowledge and institutions

Unimproved water supply/source. A source that is considered to be at risk from contamination. The WHO/UNICEF Joint Monitoring Programme (JMP) includes within this category unprotected dug wells or springs, vendor-provided water, surface water, tanker-truck supply, and, in some cases, bottled water

Vulnerability. The exposure and sensitivity of a system (or population) to external shocks and stresses, such as climate impacts, mitigated by the ability of that system to adapt

Water, Sanitation and Hygiene Committee (WASHCO). A committee nominated by a community to operate local water systems and carry out minor repairs

Water scarcity. Lack of an acceptable quantity and quality of water for health, livelihoods, ecosystems and/or production. Sometimes described as *physical scarcity*, where water availability is limiting, or *economic scarcity*, where access to water is constrained

Water security. The availability of an adequate quantity and quality of water for health, livelihoods, ecosystems and production, and the capacity to access it, coupled with an acceptable level of water-related risks to people and environments, and the capacity to manage those risks

Water service. The quantity, quality, reliability and cost of water accessible to users over time

Woreda. Administrative area equivalent to a district

Zone. An intermediary administrative unit composed of several *woredas*, usually without financial autonomy

Introduction

Roger Calow, Alan Nicol and Zemede Abebe

Achieving water security remains an elusive goal across large swathes of sub-Saharan Africa (SSA), despite the continent's relatively rich endowment of water. Why is this and what can be done? This book, an outcome of the five-year, Department for International Development (DFID)-funded Research-inspired Policy and Practice Learning in Ethiopia and the Nile Region (RiPPLE) Programme[1] addresses these questions from an Ethiopian perspective, with eight chapters covering different elements of the water security equation. In keeping with current thinking on the complex nature of water security, the book draws on a variety of disciplinary perspectives, with contributions from RiPPLE's Ethiopian partners as well as its European collaborators. Based on extensive field research, *Achieving Water Security* draws conclusions for policy and practice of relevance not just for Ethiopia, but for SSA more widely as governments across the continent seek to extend access to secure water and sanitation – and the benefits that flow from it – against a backdrop of rapid population growth, hydrological variability, and longer-term climate change.

Summary and seven propositions

Since its inception in 2006, the RiPPLE programme has generated an impressive body of knowledge on water, sanitation and hygiene (WASH) options and choices, financing, service sustainability, monitoring, and the drivers of change – including climate variability and change – that affect the ability of poor people to access reliable, affordable services. Conceived as an action-research programme, its objective was to strengthen the evidence base and support learning on WASH financing, delivery, and sustainability. At the heart of the RiPPLE approach was the proposition that research should be conducted with, rather than for, the people most likely to use it.

But why the focus on water *security* in this book, a term now widely used but rarely associated with services and sustainability? The association here is deliberate: we argue that achieving water security and building resilience is, first and foremost, about extending sustainable and affordable access to basic services. In so doing, we also contend that the water scarcity experienced by some 46 million unserved Ethiopians, and some 344 million Africans (WHO/UNICEF, 2012), has its root causes in governance, financing, and access, rather than water availability. This is not to suggest that water availability has no bearing on the ability to meet water and sanitation targets, or more particularly on the ability of governments to mobilize the large volumes of water needed to grow food, generate energy, and maintain ecosystems. Indeed rising levels of investment in irrigation and power generation – arguably long overdue – raise concerns about resource sustainability and water futures across Africa. Rather, the aim is to highlight the vital role *secure water* for basic needs plays, directly and indirectly, in poverty alleviation. Households benefit through a range of health, educational, nutritional, and broader livelihood impacts; local, regional, and national economies benefit from greater economic activity, spending, and investment; and over the longer term, households and economies benefit through greater resilience to climate change and other pressures.

This book deals with rural WASH, though the principal focus is on water services provided from groundwater-based sources – wells and boreholes. As African governments re-double their efforts to meet drinking water targets (see below), we note that the biggest challenges lie in meeting dispersed, rural demand. Although Africa has the highest rates of urban population growth, posing new challenges for service providers, Africa's population is still predominantly rural, and poverty remains an overwhelmingly rural phenomenon. In Ethiopia, roughly 85 per cent of the population still resides in rural areas. Meanwhile, Africa's population is increasing at 2.3 per cent per year, over double the rate of growth in Asia (UN-DESA, 2010).

So why have governments and their development partners struggled to meet water and sanitation targets? And how can policies and plans for delivering and sustaining services for poor people be strengthened in the face of multiple pressures? Before turning to the Ethiopian experience, we note that the arithmetic is certainly not all gloomy. Indeed the international target for halving the number of people without access to safe water – Millennium Development Goal (MDG) 7 – has already been met, five years before the 2015 deadline (WHO/UNICEF, 2012). However, the global figures are skewed by rapid progress in India and China[2] and major regional and national differences remain. Although 322 million Africans have gained access to safe water over the period 1990–2010, 65 million *more* people in Africa lacked access to an improved source in 2010 than did in 1990. Moreover, the numbers conceal major national and local disparities, particularly the divide between urban and rural populations in terms of the services provided. Progress on sanitation, meanwhile, continues to lag: over 2.5 billion people globally still lack access – over one third of the world's population – and more people live without access to sanitation today than in 1990, including 197 million additional Africans (ibid).

Surprisingly, total aid for all aspects of water as measured by the Organisation for Economic Co-operation and Development (OECD) fell from eight per cent to roughly five per cent of total overseas development assistance (ODA) between 1997 and 2008, despite ample evidence of WASH's multiplier effects on other MDGs (GLAAS, 2010).[3] Moreover, less than half of the funding for WASH from external support agencies is going to low income countries, and only a small proportion of these funds are allocated to the provision of basic services (ibid). WASH also remains a relatively low priority for domestic allocations of public spending in SSA (Banergee and Morella, 2011). In this respect, Ethiopia bucks the trend: current levels of funding for rural water supply have, at least in nominal terms, reached the annual volumes required to achieve near universal access by 2015 (AMCOW, 2011) under the Ethiopian Government's Universal Access Plan (UAP). Indeed one of the biggest challenges facing Ethiopia is the ability to *productively absorb* large and sustained budgets earmarked for WASH because of weak institutions and difficulties in harmonizing and aligning different financing modalities with core government systems (ibid*).*

As Chapter 1 notes, these donor-related bottlenecks are now being addressed through a new WASH Implementation Framework – the WIF (FDRE, 2011) – which proposes a single financing modality channelled through the Ministry of Finance and Economic Development (MoFED). However, low expenditure rates also point to more entrenched capacity problems, particularly at local levels where staff vacancies/turnover and limited recurrent (vs. investment) budgets limit the ability of water offices to deliver against targets and backstop existing infrastructure. This suggests a need to focus less on aggregate aid volumes and the appropriate level of donor generosity, and much more on the scale and direction of external support that countries like Ethiopia can productively absorb. Recent emphasis on low-cost technologies, including self-supply, may reduce the burden on local government, but not remove it altogether, as Chapter 3 makes clear. The RiPPLE-inspired Guided Learning on Water and Sanitation (GLoWS) initiative, discussed further in Chapter 8, provides an excellent example of how to build and sustain local capacity through on-the-job training, delivered through Ethiopia's Technical and Vocational Education and Training Colleges (TVETCs).

Local capacity constraints and limited operational budgets also affect the ability to sustain services over time. Achieving enduring increases in coverage that reach the poorest people and ensure hard-won public health and poverty alleviation benefits are not lost as people revert to unimproved services, remains a huge challenge throughout SSA. This is not least because so many systems fail to provide water on a continuous basis – for *security of supply* – because they fail, either intermittently or permanently. Recent figures for 20 countries in SSA suggest that roughly 35–40 per cent of handpumps are 'non-functional', representing a total investment of some US$1.2–1.5 billion over the last 20 years (Baumann, 2009; RWSN, 2010). We call this a largely hidden crisis because few detailed, rigorous studies have been carried out on this topic, and because severe limitations in monitoring and evaluation (M&E) make it very difficult to assess sector progress and outcomes. However, Ethiopia is now addressing this problem through its National WASH Inventory (NWI), described further in Chapter 2. The findings generated through the NWI – now being scaled up across the country – may make surprising reading: provider-based figures currently used to estimate access to water in rural areas, based on assumed functionality and people served, put coverage at close to 66 per cent; most commentators, and the Joint Monitoring Programme (JMP) for water supply and sanitation, suggest a figure closer to 40 per cent.

As noted in Chapter 5, the adequacy of financial flows to cover the recurrent costs needed to ensure sustainability is key, and the expectation that communities can and should meet these in full is probably misguided. This suggests a need for alternative cost sharing agreements that are not simply based on individual communities' responsibilities for their own water points, at least for rural handpump and motorized services (Carter et al., 2010). More generally, it suggests that the community-centric model of service provision that has dominated practice for two decades or more in SSA, and in which infrastructure is handed over to willing communities, needs to be re-visited. End users need to be fully involved in the planning and implementation process. However, shouldering responsibility for *all* ongoing maintenance is a tall order when the full life-cycle costs of repair and maintenance are perhaps seven times the amount typically collected in user fees (Baumann, 2009), and real incomes are static or declining. The Ethiopian Government has responded by putting more emphasis on lower-cost technologies including 'facilitated self-supply' (Chapter 3), and by asking households, communities and the embryonic private sector to do more in terms of articulating demand, planning and implementing projects, procuring goods and services, and managing funds. However, it is too early to tell whether this strategic shift will change the sustainability picture over the long term.

What of climate change and the sustainability challenge? Is this a game-changer in terms of the planning and delivery of sustainable services, or should the emphasis remain on addressing underlying vulnerabilities through existing policies and technologies? The risks posed by climate change to SSA are well rehearsed. Worth highlighting, however, is that climate change is both a significant threat to and a huge opportunity for development (DFID, 2011).

For Ethiopia, the threat posed by climate change to water resources and water-dependent services is difficult to gainsay, not least because of the difficulties of downscaling rainfall projections and 'downstream' effects on water resources. What is clear is that the country will experience a more unpredictable and variable climate with more droughts and floods – extreme events that have a history of catalysing humanitarian crises (McSweeney et al., 2008). This will impact on Ethiopia's growth and livelihoods, though less so as the economy diversifies and the importance of rainfed agriculture diminishes. The threat to WASH, however, can be over-played, at least in terms of resource sustainability, as Chapters 5 and 7 make clear. The key point here is that rural Ethiopians depend overwhelmingly on groundwater resources that offer a natural buffer against climate variability. Extending access to the natural storage that groundwater provides therefore remains key, and

WASH will continue to act as a 'first line of defence' in reducing the vulnerability of poor people to both climate variability and change (Howard et al., 2010; Calow et al., 2011). That said, climate change does pose a heightened threat to particular infrastructure types: unimproved sources and hand-dug wells tapping limited storage may dry up during extended droughts; floods can cause catastrophic damage to infrastructure; and floods can also spread excreta and contaminate water points, with obvious health risks (ibid).

In terms of opportunities, there is the promise of additional finance for the country, either from the climate funds of developed countries or from the private sector through carbon trading. Ethiopia also plans to harness its hydropower potential both to meet its own energy needs and to sell to neighbouring countries. The government's Climate-Resilient Green Economy (CRGE) initiative sets the ambitious target of developing an economy that is both climate resilient and carbon neutral by 2025 (FDRE, 2011). The opportunity here is to invest in the basic hydraulic and institutional infrastructure of water storage, distribution, delivery, and access that Ethiopia so badly needs. While the investment mix includes hydropower and irrigation to generate green energy and boost agricultural production, it should also be directed towards strengthening the delivery and sustainability of WASH services, and towards water resources management as overall water withdrawals increase. In this respect, the convoluted rhetoric on adaptation risks emphasizing activities that distract from more urgent development priorities that address existing vulnerabilities. As the World Bank (2010) notes, adaptation to climate change should start with developmental measures that tackle existing problems, with poverty alleviation as key.

Before turning to individual chapters, we conclude by stating that WASH in Ethiopia is, in many respects, a real success story *so far*. Progress under the UAP has been dramatic in terms of planning, financing, and coverage outcomes considering implementation only began in 2006. The RiPPLE programme has made a small but telling contribution, filling research gaps but also stimulating demand for further research and the evidence needed to inform policy and planning. This evidence base is crucial: Ethiopia's ability to build more resilient livelihoods is intimately linked to its ability to extend and maintain water and sanitation services, and better informed strategies and policies are central. Building on the knowledge already provided through RiPPLE, we make the following seven propositions:

- That **extending water and sanitation services is a precondition for tackling poverty and reducing vulnerability.** Secure water and sanitation bring a multitude of health benefits for poor and vulnerable people, and also contribute towards wider livelihood and economic goals, many of which are particularly relevant for women and girls.
- That the challenge of **sustaining services** will require a change in mindset and policies on the part of government, donors, and NGOs. The campaign mode approach to WASH that privileges capital investment

in new infrastructure over recurrent spend needs to change, with much greater emphasis on the maintenance and rehabilitation of existing services and the long-term support needs of communities and households.
- That an investment- and target-led approach to WASH also pays insufficient attention to the ability of institutions, particularly at a local level, to **productively absorb investment**, particularly when skewed towards capital items. Building the capacity of local government offices and addressing problems of staff skills and retention are vital long-term goals yet receive only ad hoc, isolated support from donors and NGOs. The GLoWS initiative, inspired by RiPPLE and discussed further in Chapter 8, demonstrates how this can be achieved through on-the-job training in vocational colleges.
- That the **M&E** of WASH outcomes and results requires further strengthening, as most sector stakeholders continue to focus on inputs (money) and outputs (facilities), rather than outcomes (access, equity) and impacts (health and wider benefits). Ethiopia's NWI is an important first step towards assessing outcomes, at least in terms of simple functionality and access, but will not provide information on the causes of service failure.
- That a diverse range of **service options and choices** (vs. a 'one size fits all' approach), tailored to different social, economic and environmental conditions and the capacity of local government, is more likely to meet and sustain diverse rural demands. Balanced packages of interventions include self-supply and multiple-use water services (MUS), alongside more conventional '*woreda* (district) managed projects', that are better able to build on household demand for domestic and productive water. Progress is now being made, with the new WIF outlining a range of different options and modalities. However, the ability of government support agencies to make informed choices based on sound local evidence remains weak.
- That water service provision needs to be **better integrated** with food security and asset rebuilding efforts, as well as with more obvious sanitation and health interventions. Particular priorities include: (a) public works programmes which currently reach around 6–10 million poor and vulnerable people annually, and often include 'bolt-on' water conservation and/or WASH components; and (b) disaster risk reduction (DRR) strategies which have historically been geared towards saving lives through food aid, rather than protecting livelihoods through mixed interventions that include WASH.
- That the discussion on **climate change** response in Ethiopia – and elsewhere – needs to wean itself away from the premise that adaptation is fundamentally different from development and must be programmed separately. Eliminating poverty is central to both development and building resilience, since poverty exacerbates existing vulnerability to both climate variability and change. The immediate priority is therefore to better prepare for levels of unmitigated variability and climate risk

that Ethiopia *already* faces. Extending access to WASH remains central to achieving this goal.

Overview of chapters

Turning now to individual chapters, the book begins with an overview of Ethiopia's water resources, water policies, and water institutions (Chapter 1). These are big topics, but the authors provide a rich summary, starting with a basic conundrum: Ethiopia has a relatively generous endowment of water, yet at least 35 per cent of rural Ethiopians do not have access to safe water. Moreover, unmitigated hydrological variability is estimated to cost the economy roughly one-third of its growth potential. Ethiopia's investments to mitigate these impacts and harness its water assets for power, food production, livestock, and improvements in health and livelihoods have historically been very limited. However, the development of water resources to support 'green growth' and poverty reduction now forms a key element of government policy. The Ethiopian Government's latest economic growth and poverty reduction plan, the Growth and Transformation Plan (GTP), aims to increase access to potable water by 30 per cent, extend the area under irrigation sixfold, and increase hydropower generation by a factor of five, all by 2015.

Progress in extending WASH services to a poor and predominantly rural population has been significant. Over the last six years, the Ministry of Water and Energy (MoWE) has led a process of policy reform and the development of an ambitious plan, the UAP, aimed at achieving near-universal access to water and sanitation by 2015. Impacts on the ground have been dramatic: in 1990 coverage was estimated at roughly 11 per cent of the rural population, whereas the latest government estimates put the figure at close to 66 per cent. Although coverage estimates from the JMP are more cautious (Chapter 2), and sanitation still lags far behind (Chapters 2 and 4), the overall trajectory is clearly positive. As the authors note, progress has gone hand-in-hand with vigorous decentralization, with responsibility for delivering services and meeting national targets progressively devolved to regional, zonal, and *woreda* (district) water offices. While this more 'bottom up' approach to service delivery has obvious merits, the capacity of local water offices to plan, implement, and maintain services remains limited, despite the presence of many disparate programme-based initiatives, including RiPPLE (Chapter 3).

While progress under the UAP is widely acknowledged, the ability of the country to *sustain* progress is difficult to predict. As the authors note, several key challenges remain. First, M&E is weak, especially relative to the level of government and donor investment. Coverage data have been based on inventories of built infrastructure and therefore *assumed* levels of service, rather than on outcomes monitored post-construction, an issue explored further in Chapter 2. Second, although data are limited, the sustainability of infrastructure and therefore services is clearly a problem. The Ethiopian Government is responding in different ways – through strengthening monitoring, but also through mainstreaming innovative approaches to service delivery via the Community Development Fund (CDF), and through self-supply (see Chapter 3). Third, cross-sectoral and cross-donor coordination has proved difficult to achieve (as in most countries), although the government is re-doubling its efforts to ensure alignment and harmonization between the ministries of water, health, education and finance, and also with donors. The aim is to move away from discrete WASH projects towards a programmatic approach – a One WASH Programme with a single Consolidated WASH Account (CWA).

Meanwhile, investment in Ethiopia's water resources to meet hydropower, irrigation, and industrial goals is accelerating. While few would question the need for Ethiopia to 'harness its hydrology' to stimulate growth and reduce vulnerability (Grey and Sadoff, 2007), the scale and pace of resource development raise questions about whether new infrastructure can simultaneously deliver improved livelihoods, environmental sustainability, and broad social development. The challenge highlighted by the authors lies in building an equivalent institutional platform for water resources management so that new proposals are rooted in sound basin planning. This is a long-term challenge, and it is hoped that Ethiopia's new River Basin Organizations, established now in three basins and being extended to others, will receive the necessary support and regulatory 'teeth' to ensure this happens.

In Chapter 2, the issue of WASH sector monitoring is explored in more detail, looking specifically at the rationale for monitoring, the approach taken by different sector stakeholders at global, national, and local levels, and the challenge of using data to assess progress and inform WASH policy and delivery. The focus of the chapter is domestic water services, though the authors note that many of the issues raised also apply to the monitoring of health and sanitation.

The chapter begins by re-visiting the rationale for sector monitoring, beginning at a global level with the World Health Organization (WHO)/ United Nations Children's Fund (UNICEF) JMP. Following the Millennium Declaration in 2000 and the World Summit for Sustainable Development in 2002, the JMP has tracked global progress towards the water and sanitation goal on a country-by-country basis, using internationally agreed definitions of access, and user data gathered from household surveys. This allows the JMP to report on outcomes – services accessed and used – rather than simply outputs in terms of systems provided and assumed numbers of users. As the authors note, however, the coverage figures generated can be very different from those reported by national ministries, even though the JMP itself relies on national data. Ethiopia is a case in point with (for example), access to rural water reported as 66 per cent by MoWE and 34 per cent by JMP (figures for 2010). The difference is partly attributable to the provider vs. user distinction noted above, but also arises from the lack of recent household survey data available to the JMP and differences in the definitions adopted. Does any of this matter? In short, yes: it matters to the Ethiopian Government because they want to see the acceleration in WASH investment and coverage captured in international comparisons, and all parties (including donors) would like to have an agreed approach in place for measuring progress.

In this context, the NWI now being rolled out across the country represents an important first step towards establishing a clear and agreed sector baseline. First, it will generate both provider and user data, and hence vital information on what has been built, what actually works, and what sources of safe (and unsafe) water people are using. Second, the NWI will therefore strengthen the evidence base for government and donors on value for money in the delivery chain: from inputs (investment) and outputs (infrastructure) to outcomes (services accessed and used), particularly as donors seek a credible set of figures for tracking the use of pooled funding. Third, the NWI is supported by all relevant institutions and line ministries in Ethiopia, and uses data collection methods validated by Ethiopia's Central Statistical Agency (CSA). This increases the likelihood of a reconciliation with the JMP and any revised (post-2015) global monitoring regime. Finally, the authors highlight how data can be used to strengthen service delivery as well as track progress, since water offices need to know which areas and groups of people have access to water, where schemes are not working, and what the reasons are. Hence they conclude that while the design of the NWI has been biased towards national monitoring needs, a key constituency and use is local.

So what challenges remain? Drawing on RiPPLE experience in both the design and implementation of WASH monitoring in Ethiopia, the authors highlight the following issues:

- *The use and users of monitoring data*: experience from other countries, and from other sectors, demonstrates that transforming data into information and knowledge is not a straightforward process, and good information does not necessarily translate into better decision-making. In Ethiopia, cooperation at a federal level has been achieved, but questions remain over 'buy in' to the NWI process (and data) at regional and *woreda* levels where the data could potentially be used to strengthen service provision. An underlying issue here is that the NWI was conceived as a high-level monitoring tool rather than as a resource for local planners. This is reflected in both the questions set and in the approach to data collection and analysis adopted. Moreover, who exactly will have access to the data, and in what form, remains unclear.
- *The institutional burden*: the NWI has been a massive undertaking for a country the size of Ethiopia. It follows that data analysis and presentation pose a similarly huge challenge. Indeed the authors note that in some regions, entering household data into the new WASH Management Information System could take over two years. And what of local *woreda* water offices? *Woreda* institutions were not involved in the design and planning of the NWI, and many may struggle to engage meaningfully.

What next? While it may seem too early to ask given that NWI data are still being processed, the baseline will need to be updated to retain its value for monitoring and planning, with regular spot checking of data quality. There are opportunities here to exploit new smart-phone technologies, easing the burden of data entry and transmission. And there is also an opportunity to learn from past mistakes, and ensure that the next exercise is conceived with more than one purpose in mind.

Chapter 3, on innovative approaches for extending access to water services, describes how approaches to the delivery of water services have evolved in Ethiopia, reflecting wider changes in thinking about the appropriate role of the state and other actors in service provision. The chapter then provides a critique of the dominant 'community managed domestic supply' model, and looks at how complementary approaches are taking shape.

The authors begin by briefly charting the ideological and practical shift away from government-led service provision to a community-centric approach involving end users in planning and implementation, and community ownership of assets. In part, this reflects the poor performance of 'top-down' programmes that put in place systems local people did not want, or could not afford to maintain. It also reflects shifts in prevailing ideology – a fundamental change in beliefs about the role of the state, and about communities as active development partners rather than passive recipients of aid. One outcome is a more heterogeneous approach to service provision in which a coalition of public, private, and civil society actors are involved. Another is the expectation that communities will contribute towards capital costs in cash or kind, and assume responsibility for maintenance. A common thread is a focus on *domestic* water services, managed and owned by *communities* through WASH committees (WASHCOs).

Research carried out by the RiPPLE programme highlights both the advantages and limitations of the approach. While Ethiopia's success in extending water services to a growing rural population is widely acknowledged, thorny problems remain. Highlighted here and discussed in further detail in Chapter 5 is the sustainability issue: systems continue to fail, communities struggle with the financing and practicalities of operation and maintenance, and *woreda* water offices often lack the capacity to provide effective support.

So what can be done? The research highlighted here suggests that conventional domestic supply projects could be supplemented, or in some instances replaced, with alternative approaches that cater for different water uses and respond to household demands. One example is the multiple-use water services (MUS) approach, aimed at providing water for both domestic and productive water needs from a single source, or from different sources. The argument for MUS is, on the face of it, compelling: rural households need water for drinking, washing, and cooking, but also want water for 'productive' uses such as livestock watering, small-scale irrigation, brick-making and so on. If these needs can be met, so the argument runs, households will be better-off, and better able to meet the costs of service provision. Drawing on detailed evaluations of alternative MUS pathways in East Hararghe, Oromia Region, the authors conclude that the economic benefits are significant, outweighing the additional costs associated with upgrading single-use systems. RiPPLE's findings have proved influential: in East Hararghe, the MUS approach has moved beyond the NGO sphere and into mainstream government planning; and nationally, MUS now forms part of the option menu outlined in the UAP.

RiPPLE research has also provided much needed evidence on the potential of another approach to service provision – self-supply – in which the initiative and investment to build or improve water sources comes from individual households. The Ethiopian Government has, since 2008, prioritized lower-cost technologies and increased emphasis on self-supply to help meet coverage targets. As the authors note, however, while the practice of digging family wells in Ethiopia is well established, equipping such wells with pumps and ensuring basic sanitary protection implies a role for government (and private artisans) that differs from the standard 'community supply' model. Based on research conducted with regional and local government in Southern Nations, Nationalities, and People's Region (SNNPR), and in close collaboration with UNICEF, the authors conclude that self-supply has the potential to significantly extend water access in areas with favourable groundwater conditions, particularly where households are scattered and community-based supplies become very expensive. However, if used for drinking, some simple steps should be taken to protect water quality. Experience to date with a target-driven approach to self-supply has been mixed; indeed a key lesson is that campaigns that privilege one option, or one type of approach, are unlikely to provide services that are tailored to local conditions.

Taken together, findings on MUS and self-supply suggest that different approaches, technologies, and service options are needed to extend access to water services in different contexts. This is not surprising given the diversity of environmental conditions found in Ethiopia, the range of livelihood strategies followed by rural households, or the very different planning and support capacities of government WASH teams. The challenge lies in making informed choices, and in moving beyond a 'single source–single (domestic) service' approach: sustainable water services are more likely to be provided from a source, or sources, that meet different water needs for different communities and households.

Chapter 4 sets out the findings of two studies led by RiPPLE project staff and their local research partners on how sanitation and hygiene have been promoted under the Heath Extension Programme (HEP) in Ethiopia.

In its sustained political and institutional commitment to the HEP since 2002, the federal Ministry of Health (MoH) has distinguished itself from many of its SSA counterparts, supporting this system of primary health care which has a major preventive element, in contrast to health programmes that focus predominantly on treatment by medication. Where preventive aspects are neglected by health authorities, playing down the responsibility of the health administration to provide leadership on sanitation and hygiene, the preventive power of sanitation and hygiene to reduce mortality from diarrhoea and incidence of tropical diseases is lost.

This chapter focuses on three aspects of sanitation and hygiene – excreta disposal, water quality control, and personal hygiene – and how they are delivered as part of the HEP by salaried health extension workers (HEWs) and voluntary community health promoters (CHPs) deployed in rural communities. In developing countries, where budget constraints mean that government health services have too few employed personnel, the engagement of community members to provide basic health services has been identified as a key extension strategy (in the 1978 Declaration of Alma-Ata).

The authors describe how the first step in the RiPPLE-led research was to study the policy-making process behind the sanitation and hygiene strategy in the SNNPR. The regional government led a strong campaign of political promotion and institutional mobilization to launch and roll out the strategy, designed to encourage latrine construction and improved hygiene in line with the HEP. An interesting feature of the campaign was 'ignition' documents, prepared by the regional bureau of health. These documents cleverly combined the technical information required to support latrine construction and hygiene promotion with a communication orientation to persuade politicians, motivate civil servants, and build consensus for action by a range of stakeholders. Within this framework, the HEWs and CHPs were successful. They did not use gifts of hardware (latrine slabs), but instead applied the 'software' of promotion and facilitation to boost the number of household latrines in rural villages, in the localities surveyed by the RiPPLE researchers. While many of the traditional pit latrines were found to be basic, with some questions as to their durability, the strategy was to assist households to place themselves on the first rung of the 'sanitation ladder' (as illustrated in Figure 4.1), with subsequent, technical improvements made once household members had become accustomed to latrine use. The central philosophy of the HEP is that, if the right knowledge and skills are transferred, households can take responsibility for improving and maintaining their own health.

This 2008 study concluded that the HEWs and CHPs, as the 'frontline' health workers, had a growing significance in the success of sanitation and hygiene outreach, acting as promotional change agents. Based on the observations in SNNPR, the combination of inputs employed to encourage behaviour change is described as the 'command' aspect of the local *kebele* authorities and the technical guidance of the HEWs and CHPs, plus messaging in support. More of those inputs are required, the researchers noted, to bring about changes in handwashing and water storage. Given that water access is critical for hygiene functions such as for handwashing, this study has confirmed the need for sanitation and hygiene interventions to be coordinated with water investments, which requires collaboration between government health and water agencies.

To place the HEP in a broader context, the authors compare examples of health extension initiatives in other SSA countries – government-led programmes, designed to extend across the national territory, as compared with those of donors or NGOs, targeted to particular locations. The evidence from the experiences of health extension cited is that community health workers (CHWs) can improve access to, and coverage of, basic health services in communities. Too many large-scale programmes have, however, failed in the past because of unrealistic expectations, poor initial planning, problems of sustainability, and the difficulties of maintaining quality which has unnecessarily undermined and discredited the concept of health extension. Essential features of well-performing community health programmes are the appropriate selection of CHPs and continuing education/training, together with proper supervision, support, and sufficient resources.

Based on the findings of the second phase of RiPPLE research on sanitation and hygiene in SNNPR, these points have considerable resonance for the HEP in Ethiopia. The authors explain how the work of HEWs and CHPs (in two sample districts in SNNPR) was clearly valued by communities, while certain aspects of operation of the sanitation and hygiene elements within the HEP required strengthening. They found that more attention needed to be paid to building the capacity of HEWs and CHPs, with more training for them in knowledge and skills. HEWs were lacking supervision and support from *woreda* health offices, and, in turn, HEWs needed to provide more guidance to CHPs in planning and organizing their work. Remote *kebeles* tended to receive less attention than those close to *woreda* health offices and health centres, pointing to inconsistent geographical coverage. The approach to motivation of HEWs and CHPs needed to be reviewed: retaining the services of these 'front-line' health workers/promoters, especially the CHPs, as volunteers, was becoming more difficult.

Provision of basic equipment and more materials for information, education, and communication was one way to motivate both HEWs and CHPs. As for the way they were conducting their work, the study found that messages on sanitation and hygiene needed to be more concentrated: household visits should focus on one issue per visit, with a message specific to that subject matter, supported by relevant information materials. In this way, households could best be persuaded to adopt new practices, including with arguments on dignity and privacy, alongside messages on the health benefits of improved sanitation and hygiene. It was understood that bringing about lasting behaviour change requires substantial follow-up, so that repeat visits to households were inevitable. Finally, there were calls for more collaboration between WASH stakeholders in the region, including active support from NGOs and donors to promote sanitation and hygiene under the government-led HEP. The authors note that the observations of the HEWs and CHPs consulted in the RiPPLE study in SNNPR were discussed with the Bureau of Health by the RiPPLE Regional Coordinator.

As emphasized in the conclusions to the chapter, it is to be hoped that development partners avoid setting up parallel HEPs that by-pass government, with donors and NGOs instead targeting their contributions to supplement the resources available for supporting the HEP. As noted by the chapter authors, after the 'exemplary leadership shown by the Ethiopian Government in setting the direction and ambition of the HEP', continuation of this health extension effort will be essential for achievement of the national targets of universal access to basic sanitation and 84 per cent access to improved sanitation by 2015, as well as reduction of infant and child mortality rates.

In Chapter 5 on the Sustainability of water services in Ethiopia, the authors explore in more detail some of the sustainability challenges highlighted above, going beyond the usual 'single issue' arguments that tend to dominate the sustainability debate. Drawing on RiPPLE research findings, but also wider national and international literature, the authors discuss the technical, social, institutional, financial, and environmental factors that determine whether WASH services continue to work over time.

Why devote a separate chapter to sustainability in the first place? Because although country-wide access data from the NWI have yet to be made available, the MoWE's own estimates indicate that some 20–30 per cent of water schemes are not providing water on a continuous basis, or have failed altogether. Indeed RiPPLE's own research in Halaba Special[4] and Mirab Abaya *woredas* in SNNPR, conducted in 2007/8, indicated that 43–65 per cent of water points or schemes were non-functional – problems that *woreda* and regional water bureau were unaware of. Moreover, problems were not restricted to the more complex schemes with motorized pumps (though repair of these could take up to 12 months): in Mirab Abaya, nearly 50 per cent of water points equipped with simple handpumps were not working. Ethiopia is certainly not unique in this respect: the authors cite published findings for SSA highlighting non-functionality rates of roughly 40 per cent for handpump schemes. What is striking is that sector professionals have known about such problems for years yet, until recently, it has proved extraordinarily difficult to interest government, donors, and NGOs in a problem many would rather not acknowledge. This is despite the fact that hard-won health gains and wider poverty alleviation benefits will be lost if people revert to using poor quality sources for drinking.

Importantly, the authors also highlight issues of *seasonal access* and *equity of access*, drawing on research findings from RiPPLE's Water Economy for Livelihoods (WELS) studies in SNNPR, Oromia, and Somali regions (see Chapters 6 and 7). Detailed water audits conducted for different wealth groups in different communities revealed that poorer households often struggled to meet even minimum drinking water needs in the dry season, with access to water varying as much *within* communities as it did between them.

What about root causes? The evidence is patchy; a key problem is that few comprehensive and rigorous studies have been carried on the causes and consequences of intermittent or longer-term failures, or where these failures occur.

Piecing together the evidence from Ethiopia, however, the authors identify a number of contributory factors ranging from poor technical design and construction, to lack of spare parts and the inability of hastily convened WASHCOs (largely excluding women) to mobilize cash contributions and repair systems. Underlying issues linking finance and institutions, however, are cost recovery and cost-sharing, and the ability of *woreda* and zonal/regional water offices to provide communities with ongoing support.

At a community level, the ability of WASHCOs to collect, manage, and bank user fees is clearly important, and RiPPLE's involvement in WASHCO training illustrates the importance of basic book-keeping and technical skills (see also Chapter 8). However, even with the necessary investment in software, the ability of users to meet the full life-cycle costs of operation, maintenance, and major repair is questionable. This then raises the issue of ongoing cost-sharing between communities, government, and development partners, the role of subsidies, and the ability of local government, in particular, to undertake major repairs when budgets are geared towards capital spend on new, target-orientated infrastructure. At the same time, the authors also note that many water offices are chronically under-staffed, especially in technical roles, with little indication of how the 120,000 plus skilled and professional staff needed nationally are to be recruited – and *retained*. The RiPPLE-led GLoWS initiative, discussed further in Chapter 8, will help in this respect.

Finally, the chapter considers the environmental dimensions of source, resource, and service sustainability, looking particularly at issues of groundwater quality, availability, and change. Two naturally occurring contaminants, arsenic and fluoride, are particular health concerns, with RiPPLE studies indicating that that over 10 million people could be at risk of fluorosis. Localized falls in groundwater levels can also cause water points to dry up, or precipitate mechanical failures in others, particularly where demands are heavy (e.g. peak dry season) and groundwater storage is limited. However, the sustainability of groundwater resources is generally not in question – at least for domestic and minor productive uses. Indeed groundwater development, carefully planned, is key to extending water services that support domestic and productive uses.

How are these challenges being addressed? In part, the Ethiopian Government's recent prioritizing of lower-cost technologies and approaches, including self-supply and CMPs, are an acknowledgement of the sustainability problem. Building on households' existing experience with traditional wells, and offering households and communities a greater say in service choices, service levels and the procurement of goods and services, has the potential to both accelerate coverage and increase reliability. As the authors note, however, such approaches have yet to be rolled out and evaluated *at scale*, and the focus on shallow wells is itself limiting. Questions therefore remain about the ongoing maintenance of middle- and higher-end technologies based, for example, on boreholes, and the kind of cost-sharing and partnership arrangements that need to emerge between the key players. Much is unclear, and while the new WIF name-checks the private sector as a 'significant partner', the emergence of private-sector supply chains, and public–private operation and maintenance support units (OMSUs), remains an aspiration, and difficult to scale out across large rural areas with dispersed populations.

A key thread running through the book is the link between access to water and sanitation services and poverty reduction. In Chapter 6, 'Water for livelihood resilience, food security, and poverty reduction', the authors review the evidence in detail, exploring both the direct and indirect pathways linking access with health, food security, and broader livelihood outcomes. The chapter draws on both international evidence and RiPPLE research, and includes RiPPLE's work on smallholder irrigation and livestock water management as well as WASH.

Perhaps the most obvious and well-rehearsed impact of WASH is on people's health, and the chapter begins by reviewing the now robust public health evidence on the human costs of unsafe water and sanitation. The data are shocking. At a global level, almost 2.5 million child deaths annually, and around six per cent of the worldwide disease burden in terms of disability-adjusted life years (DALYs), can be attributed to inadequate access to water and sanitation and poor hygiene practices. Diarrhoea, and subsequent malnutrition, account for most of this disease burden. In hard monetary terms, the *benefits* of universal access to safe water and sanitation are huge – for diarrhoea alone, there are savings at a household level of $565 annually and at a national level (health budgets) of some $11 billion in treatment costs.

But what of country-specific data? Here the literature is thinner, and RiPPLE research adds new insights. While an analysis of secondary data from Ethiopia's national Welfare Monitoring Surveys (WMSs) is largely inconclusive, primary data from a RiPPLE survey of 1,500 households in Oromia Region highlight a more compelling, statistically significant link between WASH, health, and broader poverty outcomes. In particular, access to improved water services is associated with a lower incidence of diarrhoeal diseases, particularly when combined with handwashing with soap. Among those seeking treatment for such diseases, health-care costs averaged $7.50 per person – over 15 per cent of annual income for the majority of households surveyed. Scaled nationally, this would imply that Ethiopian households are spending roughly $27 million per year on treating diseases related to poor water and sanitation, with the disease burden falling disproportionately on the under-5s and the over-65s. This is not the whole story, however. RiPPLE data also highlight the role access to improved water services, associated with reduced distances to water, plays in releasing labour for off-farm employment, strongly correlated with reduced levels of poverty. Drawing on RiPPLE's additional work on household water economies (see also Chapter 7), the authors also note how peak (dry season) water collection times can conflict with farm *and* off-farm labour demands, and can also compromise the ability of households to participate in cash/food-for-work programmes.

In terms of agricultural water investments, the authors summarize the rationale for agriculture-led growth and poverty reduction in SSA, and the pressing need to raise productivity in ways that maintain, or enhance, the natural resource base and minimize greenhouse gas (GHG) emissions. The Food and Agriculture Organization (FAO), in their recent *State of the World's Land and Water Resources* flagship publication (2011), term this 'sustainable intensification'. In Ethiopia, the government has set out ambitious plans for the intensification and commercialization of agriculture under the CRGE and GTP, recognizing that a shift to higher value production depends on improved water use in conjunction with soil and water conservation. Bare statistics suggest tremendous room for growth, with Ethiopia in the top six of African countries with the greatest irrigation potential. But translating potential into investment, production, and income remains hugely challenging with thin infrastructure, thinner markets, and entrenched poverty.

Against this background RiPPLE research has examined the direct and indirect impacts of smallholder irrigation on different groups, demonstrating a strong correlation between irrigation and lower incidence, depth, and severity of income and food poverty. However, findings also strike a note of caution. First, a comparison of RiPPLE findings with those emerging from other research studies highlights major variation in production and income gains. Clearly context matters, and irrigation is in most cases only likely to be viable for cash crops or high value food crops, and where markets are accessible. This is the difference between physical and economic potential. Second, gains are not evenly distributed between wealth groups, and clearly those households with greater (and more secure) assets benefit most. And third, spill-over effects appear limited, with only minor gains reported by non-irrigating households in the RiPPLE survey. A key conclusion is that any significant scaling-up of investments in agricultural water will require a comprehensive package of measures that focus on infrastructure, markets, and institutions – as well as agricultural support services.

Finally, the chapter considers the role of water management in livestock production, arguing that researchers and policy-makers have paid insufficient attention to the topic. The need to prioritize livestock needs and livestock-based livelihoods is clear: livestock production is the principal livelihood activity of 20 million pastoralists in SSA, and remains hugely important in both pastoral and agricultural areas of Ethiopia, providing draught power, manure, milk and meat, and insurance for households. The authors note that in Ethiopia, a history of top-down interventions, such as the construction of permanent ponds and boreholes, has encouraged permanent settlement – often intentionally – and led to over-grazing, erosion, and the spread of disease. However, there are positive experiences too, and the authors highlight more recent participatory approaches to water and pasture management that work with traditional management practices, and provide or safeguard a mix of temporary and permanent access points to water.

Responding to climate variability and change, the subject of Chapter 7, is now a key priority in Ethiopia, evidenced in the Ethiopian Government's new strategy to develop a climate resilient economy, and in the prime minister's prominent role in international climate change negotiations. The chapter discusses the issues around climate, water, and livelihoods in some detail, from a review of climate change projections and uncertainties to specific recommendations for supporting water-related adaptation at different scales.

The chapter begins with the now well-rehearsed Intergovernmental Panel on Climate Change (IPCC) assertion that water is the primary medium through which climate change impacts will be felt by people, ecosystems, and economies. Yet predicting the scale, magnitude, and distribution of impacts remains extremely difficult – in Ethiopia and elsewhere in SSA. In part, this is because of uncertainties in climate modelling, and in particular the difficulties associated with projecting rainfall and down-scaling to regional and sub-regional levels – the scales relevant to decision-making. Hence while there is good confidence in temperature predictions, rainfall projections are much more problematic. However, it also reflects the complexities involved in translating changing patterns of rainfall, temperature and evapo-transpiration (ET) into changes in run-off and groundwater recharge when factors such as land use 'intervene'. What is clear is that Ethiopia will experience a continued rise in temperature and a more unpredictable and variable climate, with more droughts and floods. In a country where such extremes already occur regularly and have a history of catalysing humanitarian crises, the impacts are likely to be severe, particularly in a context of land use change – part climate driven – and a rapidly growing population.

Ethiopia's vulnerability to climate change – and the vulnerability of other poor countries in SSA – can be explained by a dependence on rainfed farming, by weak institutions, and by limited infrastructure. As noted in Chapter 1, Ethiopia's water resources are also unevenly distributed, and with little infrastructure to store and distribute water, the country cannot mitigate the impacts of variability by moving water (or food via roads) between surplus and deficit areas, or reduce the risk of flooding. At the same time, entrenched poverty, particularly in rural areas, coupled with rapid population growth and land degradation, will exacerbate impacts. These will register in broad macro-economic terms as reductions in gross domestic product (GDP), estimated by the World Bank at between 2–10 per cent per year. But they will also affect the livelihoods of millions, particularly those already struggling to cope with existing climate variability. The authors illustrate this by drawing on quantitative findings from RiPPLE research on household water security. Using a WELS framework, the research revealed that even in a normal year, households struggled to meet minimum water needs for domestic use, hygiene, and livestock watering, with poorer households worst affected because of lower labour availability and fewer options for water storage and transport. Moreover, the research highlighted 'crunch points' in the seasonal calendar when tradeoffs have to be made between water collection and income-generating/productive activities, with the result that households reverted to poorer quality sources and/or sacrificed food and income.

Turning their attention to responses, the authors draw a distinction between planned and autonomous adaptation, and highlight the difference between merely coping with risks, and adapting to them over time. Importantly, the authors note that risk management is nothing new; pastoral livelihoods, for example, are uniquely adapted to climate variability and drought, and crises only tend to occur when pastoralists' migration patterns are disrupted and they cannot access reserve pastures and water sources.

In terms of planned adaptation, the authors discuss Ethiopia's initial attempts to identify key regions, sectors, and livelihoods most vulnerable to the impacts of climate change, as well as priority actions through the National Adaptation Programme of Action (NAPA). Since its publication in 2007, the Ethiopian Government has shifted responsibility for climate change assessment and planning to the Environmental Protection Authority (EPA), and developed a much more comprehensive plan to mainstream climate change across sectors and fast-track to a green economy – the CRGE. A key pillar is the development of green energy from hydropower, and Ethiopia certainly has huge potential. While there are compelling arguments for investing in hydropower (and irrigation), the environmental balance sheet is not entirely green, however. As noted in Chapter 1, rapid development of water resources without strong institutions for managing risks and tradeoffs can undermine the resource base, and squander opportunities for *responsible* growth: growth that simultaneously delivers improved livelihoods, environmental sustainability, and social equity.

Finally, the chapter looks at options for supporting adaptation to climate change and other risks across scales, and identifies priorities around information systems (especially for DRR), groundwater development for WASH and irrigation, and mainstreaming risk management into development policy more generally. Here, the authors are right to conclude that adaptation to climate change should start with developmental measures that tackle existing problems, with poverty alleviation as key. Extending access to reliable water services through the natural storage of aquifers remains an urgent priority. As noted previously, this will mean paying much more attention to the targeting, design, maintenance, and upkeep of water services, not just the building of new ones. Irrigation development will also help buffer the effects of greater rainfall variability and strengthen income and food security, though as the authors note in Chapter 6, complementary investments in markets, supply chains, and transport will be needed to maximize returns. Moreover, the kind of 'green revolution' experienced in South Asia, based on subsidized power and intensive groundwater development, cannot be repeated across rural Ethiopia where environmental, economic, and political conditions differ, or have changed. Extending water services and irrigation

– from smallholder to commercial – will also depend on a sound information base – on water resources, water demands, risks, and uncertainties. Indeed dealing with the uncertainties of climate change is fundamental, and underscores the need for planning across sectors and investments that is robust to uncertainty – i.e. appropriate to a range of potential rainfall conditions.

The first seven chapters present the range of research RiPPLE conducted from 2005–11. Chapter 8, which reflects on the RiPPLE Approach to Sector Learning, returns to examine RiPPLE's purpose as a programme, and to reflect on its innovative approach to research. RiPPLE sought to ensure that research answered the real needs of those working in the sector, and would be used to inform policy and practice. This meant working in close partnership with government ministries (and their regional and local offices), civil society organizations (CSOs), universities, and private-sector organizations, bridging the traditional divide between research and practice, and bringing findings from the ground into policy debates.

At the heart of RiPPLE's approach to research were Learning and Practice Alliances (LPAs): stakeholder platforms established at different levels (national, regional, and *woreda*) whose members identified knowledge gaps they faced in their work and then participated in the research itself, under the leadership of experienced researchers. LPAs were established at national level (the Forum for Learning on Water and Sanitation – FLoWS), at regional/zonal level in three regions (SNNPR, Benishangul-Gumuz, and East Hararghe Zone of Oromia Region), and in two selected *woredas* within each of these. At each level, full-time coordinators were appointed to facilitate the LPAs, engaging local stakeholders and encouraging wide debate around research findings. The multi-level LPA approach meant that findings from local-level research could rapidly be fed into higher-level policy discussions, and the fact that local government staff had participated in the studies lent them credibility. At local level, too, LPA members including local government staff valued the exposure to realities on the ground and the practical nature of the research.

Involving such a range of people in research teams – some of whom had little experience of conducting formal research – brought obvious challenges. Intensive backstopping and guidance from experienced researchers was needed to ensure quality. The benefits, however, were significant. Research teams were committed, debated new evidence robustly, and in many cases have drawn on the experience to inform ongoing work. Collaboration between government, civil society, and academia – and even between different government departments – was found to be very limited at the start of the programme. When brought together in working teams, these different actors found many opportunities to learn from each other, and this chapter includes many of their comments and perspectives. Through the LPA process, RiPPLE hoped to promote more collaborative ways of working which would persist beyond the programme's lifetime, and extend beyond LPA meetings. It is too early to comment on whether this has happened, and institutionalizing more integrated practice remains a challenge, but there is evidence that the LPA process was valued highly for the learning opportunities it provided. 'RiPPLE has injected the importance of learning into the sector', noted one Ethiopian commentator.

The programme adopted various strategies to strengthen documentation and use of evidence in the sector, including the establishment of resource centres, wide dissemination of publications in hard copy, and on its website (www.rippleethiopia.org). But institutionalizing new approaches based on evidence requires more than making that evidence available. RiPPLE built a reputation which gained it a place in national policy debates, both in its own right and as a member of civil society coalitions, and enabled it to co-convene influential events on important policy questions such as the design of the National WASH Inventory and approaches to MUS. To ensure that learning from the programme was also reflected in the practices of those who really manage services on the ground, and to bolster the vital capacities of key local institutions responsible for water, RiPPLE also developed training courses for WASHCOs and *woreda* water officers. The latter is now being institutionalized through TVETCs nationwide, and will provide trainees with valuable practical experience to complement the existing classroom-based training.

Through these various approaches, RiPPLE targeted critical knowledge, learning, and capacity gaps to increase the prospects of ambitious sector goals becoming a reality. Perhaps more importantly, it sought to foster more collaborative, evidence-based ways of working in a country where central policy narratives are often promoted with limited opportunity for bottom-up learning. And RiPPLE is continuing this important work, now as an independent NGO registered with the Ministry of Justice. The authors of this introductory chapter – all former Directors of the RiPPLE programme – wish it every success in the years ahead.

Notes

1. DFID funding for the RiPPLE programme ended in June 2011. In November 2011 RiPPLE became an independent Ethiopian NGO.
2. These two countries account for 47 per cent of the 1.8 billion people who gained access to improved water and 38 per cent of the 1.3 billion people who gained access to improved sanitation between 1990 and 2008.
3. Furthermore investment in WASH has not kept pace with increases in other basic services such as education and health.
4. The locations Hawassa, Halaba and Haramaya were formerly known as Awassa, Alaba and Alemaya respectively.

References

African Ministers' Council on Water (AMCOW) (2011) 'Water supply and sanitation in Ethiopia: Turning finance into services for 2015 and beyond', An African Ministers' Council on Water Country Status Overview, Water and Sanitation Program – Africa region, World Bank, Nairobi, Kenya.

Banergee, S.G. and Morella, E. (2011) *Africa's Water and Sanitation Infrastructure: Access, Affordability and Alternatives*, World Bank Directions in Development – Infrastructure, World Bank, Washington D.C.

Baumann E. (2009) 'May-day! May-day! Our handpumps are not working!', Rural Water Supply Network Perspectives No. 1, RWSN Secretariat, St-Gallen, Switzerland.

Carter, R.C., Harvey, E. and Casey, V. (2010) 'User financing of rural handpump water services', Paper presented at the IRC Symposium Pumps, Pipes and Promises, IRC, The Netherlands.

Derpartment for International Development (DFID) (2011). *Climate Briefing: What is the UK Doing to Support Ethiopia's Response to Climate Change?*, DFID, UK

Food and Agriculture Organization (FAO) (2011) *The State of the World's Land and Water Resources for Food and Agriculture (SOLAW) – Managing Systems at Risk*, FAO, Rome and Earthscan, London.

Federal Democratic Republic of Ethiopia (FDRE) (2011) Ethiopia's Climate-resilient Green Economy: Green Economy Strategy, FDRE, Addis Ababa.

FDRE (2011b) *The WASH Implementation Framework (WIF) – Summary*, version: 27 July 2011, FDRE, Addis Ababa.

Global Analysis and Assessment of Sanitation and Drinking-Water (GLAAS) (2012) *UN-Water Global Analysis and Assessment of Sanitation and Drinking-Water Report: The Challenge of Extending and Sustaining Services*, WHO, Geneva.

Grey, D. and Sadoff, C. (2007). 'Sink or swim? Water security for growth and development', *Water Policy*, 9: 545–71 <http://dx.doi.org/10.2166/wp.2007.021> [accessed July 2012]..

McSweeney, C., New, M., and Lizcano, G. (2008) *UNDP Climate Change Profiles – Ethiopia* [website]. Available from: <http://country-profiles.geog.ox.ac.uk> [accessed July 2012].

Rural Water Supply Network (RWSN) (2010) 'Myths of the rural water supply sector', Rural Water Supply Network Executive Steering Committee, Perspectives No. 4, RWSN Secretariat, St-Gallen, Switzerland.

United Nations Department of Economic and Social Affairs (UN-DESA) (2010) *World Population Prospects, the 2010 Revision* [website], United Nations, Department of Economic and Social Affairs, Population Division, New York. Available from: <http://esa.un.org/unpd/wpp/index.htm> [accessed July 2012].

World Bank (2010) *Economics of Adaptation to Climate Change – Social Synthesis Report*. The World Bank, Washington D.C.

World Health Organization (WHO)/United Nations Children's Fund (UNICEF) (2012) *Progress on Drinking Water and Sanitation: 2012 Update*, WHO Press, Geneva.

WHO (1978) *Declaration of Alma-Ata* [online], WHO, Geneva. Available from: <www.euro.who.int/__data/assets/pdf_file/0009/113877/E93944.pdf> [accessed July 2012].

About the authors

Roger Calow is Head of the Water Policy Programme at the Overseas Development Institute (ODI) and an Honorary Research Associate at the British Geological Survey. He has over 20 years' experience leading international research and development projects in Asia, Africa, and the Middle East, including two years as Director of the RiPPLE programme in Ethiopia. He leads an interdisciplinary team of eight staff working on water supply and sanitation, climate change and water security, water resources management, and the political economy of sector reform.

Alan Nicol is a Research Fellow in the Knowledge, Technology and Society (KNOTS) team at the Institute of Development Studies (IDS). He specializes in water and sanitation and water resources management, and has over a decade of professional experience leading policy-related research programmes. This included three years in Ethiopia, where he established and directed the RiPPLE Programme from 2006 to 2009. His major fields of interest are water resources development, climate resilience and adaptation, the political economy of policy development, and the links between water, livelihoods and poverty reduction. He works at many different scales from community to transboundary basin and in recent years has focused on regional integration issues in the context of shared river basin development on the Nile and in south Asia.

Zemede Abebe is a Programme Advisor for RiPPLE and the Hararghe Catholic Secretariat in Ethiopia, and is also leading the Disaster Risk Reduction (DRR) Team for the Catholic Relief Service (CRS) in Eastern Africa/South Sudan. With expertise ranging from water resources management, to agro-enterprise and building learning alliances, Zemede has over 15 years' experience working in a range of research, emergency, rehabilitation, and development programmes, and was national Director of RiPPLE from 2009 to 2011. He holds an MSc in Agricultural Economics from Haramaya University, Ethiopia.

CHAPTER 1
Ethiopia's water resources, policies, and institutions

Eva Ludi, Bethel Terefe, Roger Calow and Gulilat Birhane

Ethiopia's economy is growing rapidly, but the country still has a poor, fast growing, and largely rural population heavily dependent on rainfed agriculture. Although Ethiopia has abundant water resources, they are distributed unevenly between areas, and hydrological variability is estimated to cost the country a third of its growth potential. Ethiopia's investments in mitigating these impacts and developing its water resources for power, food production, livestock, and improvements in health and livelihoods have historically been very limited. However, the Ethiopian Government's latest economic growth and poverty-reduction strategy, the Growth and Transformation Plan (GTP), aims to reverse the trend through significant investment in hydropower, irrigation, and flood control. Over the last six years, the Ministry of Water and Energy (MoWE) has also invested in an ambitious initiative – the Universal Access Plan (UAP) – aimed at achieving near-universal access to water and sanitation by 2015. Impacts on the ground have been dramatic and progress has been supported through a vigorous decentralization process, but major challenges remain, particularly in ensuring services are sustainable, and in building the capacity of local water offices to plan projects and provide ongoing support services for households and communities. As water withdrawals for irrigation, hydropower, industry, and urban consumers increase, parallel investment in water resources management and monitoring is required. Of particular importance for embryonic river basin organizations are legal frameworks, rights administration, and the development of robust and flexible allocation mechanisms. Ethiopia's Water Resources Management Policy and Water Sector Strategy provide the basic building blocks, but may need to be updated to account for emerging concerns around climate change, disaster risk reduction (DRR), and the protection of ecosystem services.

Ethiopia's water resources

Compared with many countries in sub-Saharan Africa (SSA), Ethiopia has a relatively generous endowment of water, with a mean total surface water flow of roughly 122 billion m^3/year, and renewable groundwater resources estimated at 2.6 billion m^3. A common indicator of scarcity is water availability per capita, estimated at country or basin level. For Ethiopia the figure is roughly 1,900m³/capita/year, placing the country above the commonly used 'water scarcity threshold' of 1,000m³/capita/year (Falkenmark, 1989). However, this is a crude indicator of physical water availability that does not consider either temporal and spatial variability, or people's access to water – 'economic scarcity' – mediated through infrastructure and institutions (CAWMA, 2007).

Ethiopia's hydrology is influenced significantly by its topography. Ethiopia is a dome-shaped country with a central highland plateau surrounded by lowlands and dissected by deep ravines. The Great East African Rift Valley divides the country into the west and east *massifs*. The highlands receive relatively high rainfall, with run-off flowing in different directions to the surrounding lowlands and, in many cases, crossing international boundaries; hence the country's label as the 'Water Tower of East Africa' (Nuru, 2012). No rivers flow into Ethiopia from neighbouring countries.

Ethiopia can be divided into eight river basins, one lake basin, and three dry basins that do not support any perennial rivers (Figure 1.1). According to Awulachew et al. (2007), these basins can be categorized as follows: river basins (Tekeze, Abbay, Baro–Akobo, Omo–Gibe, Genale Dawa, Wabi Shebele, Awash, Denakil); lake basin (Rift Valley Lakes); dry basins (Mereb, Ayisha, Ogaden). With the exception of the Awash River and Rift Valley Lakes Basins, these are transboundary. The Abbay, Baro–Akobo, Mereb, and Tekeze Rivers flow into Sudan, cross into Egypt, and drain to the Mediterranean, forming part of the Nile Basin system. The Omo–Gibe River is the major tributary to Lake Turkana, which lies between Ethiopia and Kenya. The Omo–Gibe enters the Ethiopian part of Lake Turkana, making the lake an international water basin. The Genale Dawa and Wabi Shebele Rivers flow into Somalia before disappearing into the sand near the Indian Ocean. The remaining three basins are also transboundary, although they do not generate any transboundary run-off (Nuru, 2012).

The occurrence of groundwater in Ethiopia is influenced by the country's geology, geomorphology, tectonics, and climate. These factors influence the availability, storage, quality, and accessibility of groundwater in different parts of the country. In some lowland areas (e.g. Somali Region) groundwater is only available at depth. In other areas, its quality poses a risk to human health (e.g. from high fluoride concentrations in the Rift Valley). Across much of the country, however, groundwater is of potable quality and can be developed in a cost-effective manner to meet dispersed demands. Hence groundwater, accessed through wells, boreholes or springs, probably provides over 90 per cent of improved rural water supply and underpins efforts to achieve the drinking water targets set out in the UAP.

Ethiopia's water resources are characterized by very high spatial and temporal variability. The western region, with relatively high rainfall, includes the four basins with the most water: the Abbay (Blue Nile), Baro–Akobo, Omo–Gibe, and Tekeze. The remaining eight basins in the central and eastern parts of the country face water shortages (Nuru, 2012). Most rivers

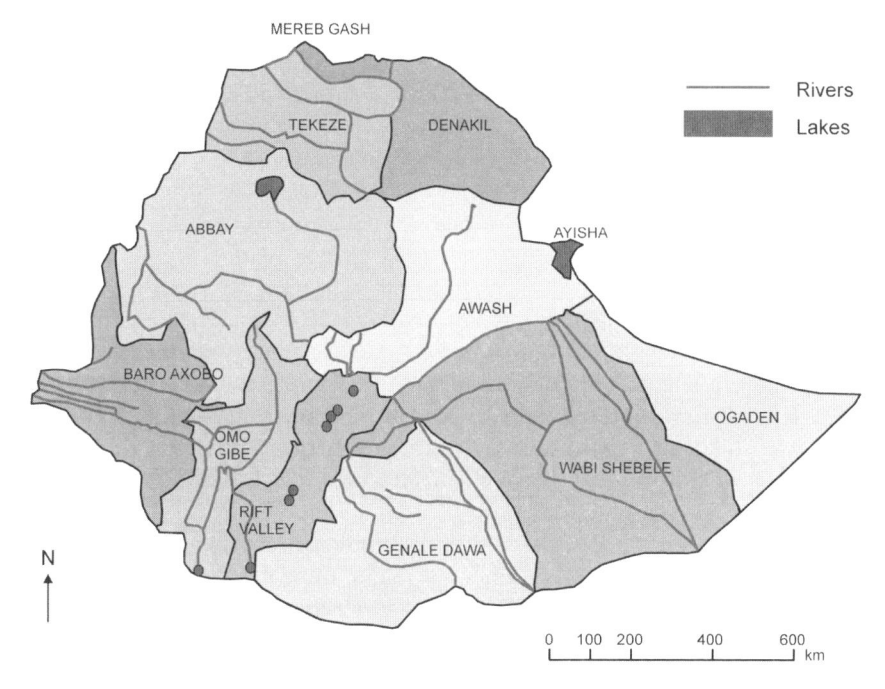

Figure 1.1   Schematic map of Ethiopia's river basins

Source: Nuru, 2012

discharge the bulk of their flow during the four rainy months from June/July to September/October, with baseflow from groundwater supporting flows over the remainder of the year. However, some rivers dry up completely for a few months before the onset of the rains. Moreover, because rainfall across much of the country is both highly seasonal and variable, droughts and floods are endemic and likely to increase in frequency and intensity as climate change accelerates over the coming decades (World Bank, 2006).

The availability of water with respect to population distribution and settlement also presents challenges. Roughly 85 per cent of Ethiopia's surface water is found in the western basins, but only 40 per cent of the population live in these areas. The bulk of the population is concentrated in the highlands because of favourable climatic conditions, but water storage in these areas is lower. The lowlands have greater surface water flows, groundwater storage, and land availability, but remain sparsely populated.

Access to water supply and sanitation

Water supply coverage in Ethiopia has increased significantly over the last two decades. According to government data, water supply coverage has risen from 19 per cent in 1990 (11 per cent rural, 70 per cent urban) to around

Pumps drawing irrigation water from the Awash River

Source: Eva Ludi

69 per cent in 2010 (66 per cent rural, 92 per cent urban), suggesting that Ethiopia has already met the Millennium Development Goal (MDG) drinking water target, to reduce by half the numbers of people without access to safe water (MoFED, 2010). For sanitation coverage, the figures are less impressive. Nonetheless, government data again show progress despite rapid population growth, from a baseline of close to zero in 1990 to around 60 per cent coverage by 2010, although only 20 per cent of households use latrines (MoH, 2010; MoH, 2011).

Coverage estimates from the Joint Monitoring Programme (JMP) are significantly lower, with access to drinking water reported at 44 per cent in 2010, 34 per cent rural and 97 per cent urban (WHO/UNICEF, 2012). Reasons are discussed further in Chapter 2, but include a lack of recent household surveys for JMP estimates, different definitions of access, and weaknesses in national monitoring systems resulting in a lack of verifiable data. However, JMP data still highlight major progress in extending coverage, with over 22 million people gaining access to drinking water between 1990 and 2010 – the second highest increase in SSA (after Nigeria). Access to improved sanitation facilities has also increased according to the JMP – from three per cent in 1990 to 21 per cent in 2010 (19 per cent rural, 29 per cent urban). However, almost 50 per cent of the population still resort to open defecation (WHO and UNICEF, 2012).

The Ethiopian Government set its own ambitious targets in the first UAP in 2005 which set out to reach full access to water and sanitation by 2012. The UAP has since been revised and aligned with the GTP (see below), with dates extended to 2015 and targets reduced to 98.5 per cent coverage (MoFED, 2010).

While improvements in access to water and sanitation are difficult to translate directly into measurable socio-economic outcomes, they are likely

WATER RESOURCES, POLICIES, AND INSTITUTIONS   29

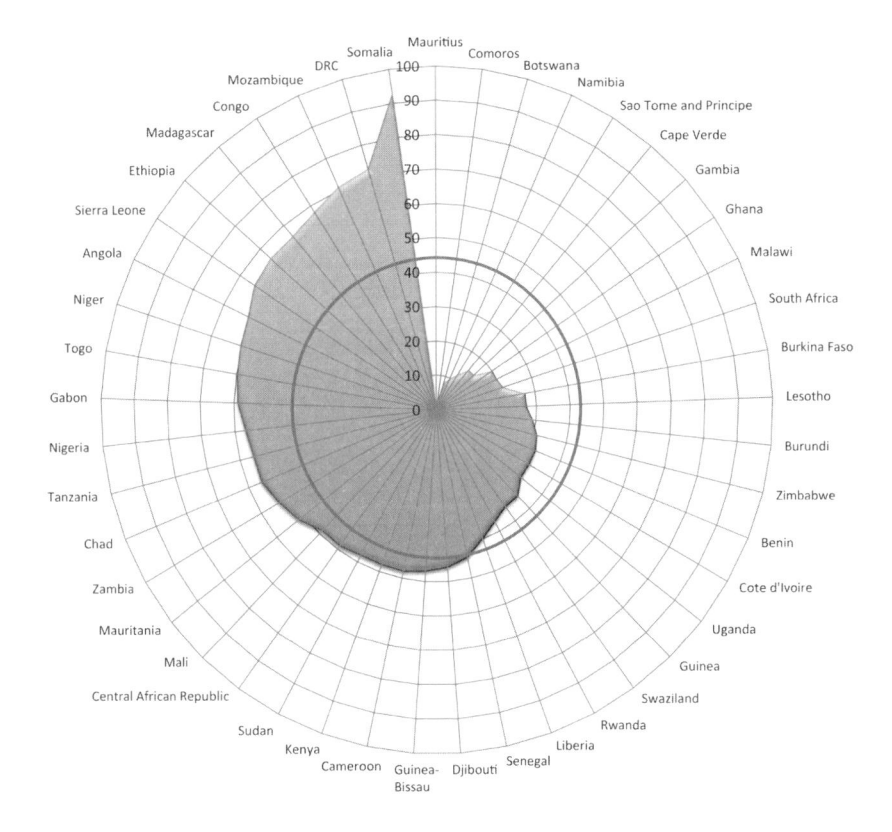

Figure 1.2   Proportion of national populations without access to improved drinking water

Note: The inner circle in bold denotes average for SSA (42%)

Source: WHO and UNICEF, 2012

to carry substantial benefits. These include time saved, better health, more girls in school because they no longer have to spend time collecting water, increased security for women once exposed to violence while collecting water, and livelihood security, better nutrition, and higher income (Hutton and Haller, 2004; Slaymaker et al., 2007).

Overview of key sector policies

Improved access to water supply and sanitation is a key indicator of social development and forms a major part of most countries' poverty reduction strategies. In Ethiopia, the overarching strategy is the GTP (MoFED, 2010), which sets out a national development path for the period 2010/11 to 2014/15.

The GTP recognizes the importance of water provision and the development of relevant institutions to manage water service delivery at appropriate administrative levels. It includes targets for both urban and rural water supply and sanitation, as well as irrigation development and electricity (from hydropower) production (Table 1.1). The GTP states that the key objective for the water sector from 2010–15 is 'to develop and utilize water for different social and economic priorities in a sustainable and equitable way, to increase the water supply coverage, and to develop irrigation schemes so as to ensure food security, to supply raw materials for agro-industries and to increase foreign currency earnings' (MoFED, 2010: 41).

As agriculture will remain a key sector for economic growth and livelihoods in Ethiopia, the GTP includes a separate section on agricultural water, focusing on the appropriate use of rainwater and an expansion of irrigation. It explains that in areas where groundwater is easily accessed, farmers will be supported to develop hand-dug wells for home gardens, vegetable growing, and permanent crop production. According to the GTP, Ethiopia has 5.1 million hectares of land suitable for irrigation development through various technologies and approaches, including using foreign direct investment. Supporting small-scale irrigation schemes, which should expand areas under double cropping, will be accelerated. A further section of the GTP deals with water in pastoral areas, highlighting the need for water resources development for both livestock and humans, in conjunction with pasture improvement and irrigation scheme development. The plan also estimates that Ethiopia has the potential to generate 45,000 megawatts (MWs) of energy through hydropower, and aims to increase production from 2,000 to 10,000 MW by 2014/15.

Table 1.1 Water-related targets in the Growth and Transformation Plan

	Baseline 2009/2010	*Target 2014/2015*
Potable water coverage (%)	68.5	98.5
Urban potable water coverage (source within 0.5 km)	91.5	100
Rural potable water coverage (source within 1.5 km)	65.8	98
Reduce non-functional rural water supply schemes (%)	20	10
Developed irrigable land (%)	2.5	15.6
Power-generating capacity (MW)	2,000	10,000

Source: MoFED, 2010

Overall, the GTP highlights the need for integrated and sustainable development and use of water resources for multiple purposes by linking different sectors and users, while ensuring equitable use of resources at basin level. The plan notes the need to use water for maximum social and economic development, and to mitigate the impacts of run-off, drought and other natural hazards. It plans for both increased access to water and for water's increased contribution to food security and growth, and recognizes the importance of watershed management and water/moisture conservation to help Ethiopia cope with climate change.

Ethiopia's National Adaptation Programme of Action (NAPA) also highlights the role of water development and management in alleviating poverty and building resilience to climate change, including a number of water-related priority actions (see Chapter 7).

Both the GTP's and NAPA's water-related priorities are guided by Ethiopia's Water Resources Management Policy (MoWR, 1999) and a Water Sector Strategy (MoWR, 2001) that emphasize the need for efficient, equitable, and optimum utilization of available water to achieve sustainable socio-economic development (Box 1.1). These documents were developed some time ago, and may need to be revised in light of more recent knowledge of climate change, and more recent thinking on Integrated Water Resources Management (IWRM) that stresses ecosystem services as well as direct human needs.

Box 1.1 Ethiopian Water Resources Management Policy (1999)

Ethiopia's Water Resources Management Policy is based on the following principles:

- Ethiopian citizens shall have access to sufficient water of acceptable quality to satisfy basic human needs. The policy prioritizes drinking water over other uses, but recognizes that water is both an economic and social good.
- Water resources development should be based on decentralized management and participatory approaches. Management of water resources shall include all stakeholders, including the private sector, and ensure social equity, system reliability, and sustainability.
- IWRM is emphasized: the policy recognizes the hydrologic boundary or basin as the fundamental planning unit and water resources management domain. Ownership is developed to lower tiers and management autonomy is at the lowest administrative level.
- Full cost recovery is the aim for urban water supply systems and recovery of operational and maintenance costs for rural schemes.

Source: MoWR, 1999

In addition to extending the target date for achieving (near) universal access to water, the revised UAP also endorses a shift towards lower-cost technologies. This includes 'self-supply', in which the initiative and investment to build or improve water sources such as family wells comes from individual households (see Chapter 2). The revised UAP also highlights the need for community mass mobilization, advocacy and promotion required to achieve the UAP targets, and the development of minimum capacity at *woreda* (district) level to implement projects, or to promote and support self-supply (MoWE, 2010).

Financing WASH services

The funding landscape for WASH in Ethiopia has a complex history. Post-war rehabilitation after 1991 was carried out through the multi-donor funded Ethiopian Social Relief, Rehabilitation and Development Fund (ESRDF), a nationwide programme that ran from 1996 to 2004 aimed at building regional capacity for project implementation. As part of wider decentralization efforts, responsibility and capacity for water-supply investments shifted from federal government to regional water bureaux. After 2004, a second wave of decentralization followed, devolving service delivery responsibility to the *woreda* level, which now allocates capital budgets to water supply (World Bank, 2009).

More recently there has been a progressive shift – albeit incomplete – from bilateral support provided outside the Ethiopian Government's core systems, to multilateral funding harmonized under a single financing window (Channel 1b, see Box 1.2) and channelled through the Ministry of Finance and Economic Development (MoFED). Hence the three largest official development partners – the World Bank, the UK Department for International Development (DFID), and the African Development Bank (AfDB) – have now joined UNICEF in harmonizing under a single financing modality. Meanwhile, most other water sector development partners, including non-governmental organizations (NGOs) and remaining bilateral donors (the European Development Bank, the Government of Finland, Japan International Cooperation Agency (JICA), and French Development Agency) support the common WASH Implementation Framework (WIF) and a jointly-agreed approach to monitoring (through the National WASH Inventory, see Chapter 2), even though they still fund discrete WASH projects. The latter may change, however; the WIF envisages a fully harmonized and aligned WASH sector – a One WASH Programme and a single Consolidated WASH Account (CWA).

One outcome has been a more joined-up approach to funding, planning, implementation, and monitoring, but one that is still plagued to some extent by incomplete harmonization and alignment. This has led to a number of problems. A study by the World Bank (2009) concluded that information regarding on-budget funding at *woreda* level (Channels 1, 1b, and 2) is generally available and can be used for planning purposes. Off-budget financing is more difficult to account for and use for planning purposes. Disbursement of funds to local levels can take time as getting money to decentralized spending units involves a series of transactions. Channel 1b, in particular, suffers from delays in channelling money to the special accounts because of reporting

> **Box 1.2 Financing modalities for WASH**
>
> Tracking budget allocation to, and spending on, WASH is notoriously difficult in Ethiopia. One reason is the continuing use of different funding mechanisms and financing modalities as summarized below:
>
> - Channel 1: includes finance managed through the government's core budget and expenditure system and allocated to regions through block grants (on-budget, on-treasury).
> - Channel 1b: a relatively new arrangement to pool donor and government funds for WASH. Money is managed by the finance ministry through a cascade of special accounts at federal, regional, and *woreda* levels (on-budget, on-treasury).
> - Channel 2: channelled through line ministries to spending units in regions or directly to regional bureaux (on-budget, off-treasury).
> - Channel 3: finance allocated by donors and civil society organizations (CSOs) directly to implementers and service providers (off-budget, off-treasury).
>
> *Source:* World Bank, 2009

challenges. Parallel accounting mechanisms, particularly for larger sums such as those related to Channel 1b and 2, can disrupt implementation. Research by the Research-inspired Policy and Practice Learning in Ethiopia and the Nile Region (RiPPLE) programme (Box 1.3) in East Hararghe Zone highlighted similar concerns.

Budget utilization rates vary from 27 per cent to 100 per cent between financing modalities. Low utilization rates are typical for modalities with their own parallel procedures, particularly for accounting but also for procurement. Indeed the African Ministerial Conference on Water (AMCOW) (2011) notes that many donor programmes have been plagued by low levels of budget utilization, particularly in the urban sub-sectors where procurement procedures are more complex.

Contrary to their influence on utilization rates, rural water supply financing modalities have limited impact on the quality and sustainability of decentralized schemes. Targeting of interventions, quality of services, and institutionalization of cost recovery were reported as satisfactory across all financing modalities (World Bank, 2009).

In terms of financing levels, estimates for required and anticipated investment based on government figures and summarized by AMCOW (2011), suggest that rural water supply is almost sufficiently well-resourced to reach the ambitious UAP target by 2015 (Table 1.2). Specifically, anticipated public investments of US$163 million per year (M/year) (from government, donors, and NGOs), plus leveraged household contributions of $73 M/year, leave a capital expenditure (CAPEX) deficit of $24 million. This has been made possible by recent budget growth from government and donors, and because of the emphasis on low-cost technologies discussed below, and in further

Box 1.3 Financing WASH services in East Hararghe: findings from RiPPLE research

Poor utilization of WASH budgets at *woreda* level has been seen in many areas, due largely to low institutional capacity and poor budget implementation and control. RiPPLE launched a study on the budgeting process in two *woredas* in East Hararghe – Babile and Goro Gutu – to assess how budgets at *woreda* level are formulated, approved, implemented, and monitored.

In both *woredas*, the major budget source is the block grant from central government, based on the revenue-collecting capacity of *woredas* and a needs-assessment report and projected socio-economic outcomes. A particular problem was inaccurate figures on socio-economic trends or needs, both leading to unfair allocation of financial resources to *woredas*. On average, from 2002 to 2007, block grants accounted for 88 per cent of the total *woreda* budget for Babile and 83 per cent for Goro Gutu: the remaining 12 per cent and 17 per cent were locally collected revenues.

There was no separate budget line for water supply before 2002, when independent water offices were established. Given that water supply is a priority poverty sector in Ethiopia, its share of the *woreda* budget is, at less than one per cent in both *woredas*, extremely low. Education, in contrast, received between 39 per cent and 41 per cent of the *woreda* budget, followed by health and agriculture with budget shares of around 14 per cent. Although the budget for water supply has increased in recent years, it is still insufficient and skewed towards recurrent expenditures at the cost of capital investments.

Water supply investments in both *woredas* were also funded through the Productive Safety Net Programme (PSNP), but this contribution could not be assessed because it is off-budget at *woreda* level.

Further investments in the water sector were hampered by the following factors:

- Until 2004/5, the water office was not represented in the *woreda* cabinet that approves budgets and so could not lobby for a larger share.
- *Woreda* officials were reluctant to allocate a higher share of budget to the water sector as they assumed the sector would benefit from direct NGO and PSNP investments.
- There was a lack of trained staff members, with, at times, only three out of nine positions filled.
- There were lengthy procurement processes.
- Underdeveloped markets meant that required materials and spare parts were not supplied.

Source: Alemu et al. (2010)

---

detail in Chapter 3. For urban water supply, however, there is a shortfall in anticipated expenditure investment, even assuming over half the total costs will be met by users (AMCOW, 2011).

In contrast there is no financing gap for household urban and rural sanitation because of the assumption that users themselves will meet the full costs ($795 M/year) of sanitation hardware. However, the AMCOW study (2011) notes that the level of investment in promotional work needed to stimulate

Table 1.2 Water supply and sanitation coverage and investment figures

	Coverage			Target	Population requiring access ('000/year)	CAPEX requirements			Anticipated public CAPEX				Assumed household CAPEX	Total deficit
	1990 (%)	2009 (%)		2015 (%)		Total	Public		Domestic	External	Total			
						(US$ million/year)								
Rural water supply	11	62		99	6,029	117	105		46	68	114		13	–
Urban water supply	70	89		99	617	143	64		3	46	49		60	34
Water supply total	19	66		99	6,646	260	169		49	114	163		73	24
Rural sanitation	4	30		99	9,363	692	0		7	23	30		692	–
Urban sanitation	25	88		99	634	102	0		19	0	19		102	–
Sanitation total	7	39		99	9,997	795	0		26	23	49		795	–

Source: AMCOW, 2011

this level of household demand and willingness to pay, is currently insufficient. Moreover, significant (unrecorded) investment will be required to meet major urban sewerage and institutional (schools and health facilities) needs.

Sustainability and functionality

The sustainability of services remains a major challenge and is discussed further in Chapter 5. Non-functionality levels reported by the Ethiopia regional water bureaux and projected by the GTP planning team (MoWE, 2010) are shown in Table 1.3.

Table 1.3 Non-functionality of rural water schemes in 10 regions

Region	Baseline non-functionality (%)
Tigray	18
Gambella	30
Benishangul-Gumuz	25
Dire Dawa	30
Harari	35
Somali	30
Amhara	18
Afar	25
Southern Nations, Nationalities, and People's Region (SNNPR)	25
Oromia	20
National average	*20*

Source: MoWE, 2010

Field studies by the World Bank (in 20 *woredas*) and RiPPLE (in two *woredas*) have found that non-functionality rates can reach 40 per cent or even 60 per cent (World Bank, 2009; Abebe and Deneke, 2008; Deneke and Abebe, 2008). Furthermore, many households classified as having access to a functional water scheme in fact travel longer distances to water and access less water per capita than the UAP service level of 15 litres per capita per day within 1.5 km (Moriarty et al., 2009). This is because access often declines in the dry season when fewer sources are available (e.g. springs), and pressure on remaining sources (e.g. boreholes) increases (Coulter et al., 2010; Tucker et al., forthcoming).

The reasons for poor sustainability may include low levels of capacity for operation and maintenance among rural water and sanitation committees (WASHCOs), inability to collect sufficient revenue from users to fund repairs, and inadequate support (e.g. for major repairs) from government agencies (see Chapter 5). In addition, the lack of reliable WASH data is a well-recognized problem, contributing to poor planning and sustainability of services. Improving the WASH monitoring and evaluation (M&E) system is now a priority, and a national WASH inventory (NWI) is being rolled out to improve information availability. The quality of inventory data and prospects for transformed sector planning are, however, limited by capacity constraints for data management and analysis at all levels, and political sensitivity around coverage figures (Butterworth et al., 2009).

Sustainability is also a major problem in sanitation. A limited supply chain and low sanitation market development means low coverage of sanitation facilities, particularly in rural areas. Although there have been waves of latrine construction in regions such as SNNPR, and improvements in awareness through the Health Extension Programme (HEP), the durability and safety of latrines and long-term behavioural change in latrine use and handwashing remain problematic (Mekonnen et al., 2008).

WASH responsibilities

Responsibility for WASH is shared by several government ministries, with MoWE leading on water supply and the Ministry of Health (MoH) on sanitation and hygiene. The MoWE (previously Ministry of Water Resources – MoWR), MoH, and Ministry of Education (MoE) have a Memorandum of Understanding (MoU) to facilitate joint planning, implementation, and monitoring of WASH to accelerate access (MoWR et al., 2005; MoWE et al., n.d.). Major areas of cooperation on paper are: water supply; human waste management; services in schools and health institutions; water quality control and surveillance; sanitation and hygiene promotion and education; personal hygiene; safe water chain maintenance; safe disposal of faeces; and solid and liquid waste disposal. Under the MoU, the three ministries committed to work together at federal level, and to support sub-national government agencies to do likewise. More recently, the MoU was extended to include the MoFED and the sector Development Assistance Group (DAG) of government, donors, and CSOs. Roles and responsibilities are outlined in Box 1.4.

Guiding principles in WASH service delivery

The WIF that accompanies the GTP and UAP (FDRE, 2011) sets out a vision for an integrated One WASH Programme, led by the Ethiopian Government. Building on the WASH MoU, the WIF is guided by principles of integration, harmonization, alignment, partnership, decentralization, and cost recovery.

Integration

WASH integration is prioritized because of the interdependence and complementarity of improved water supply, sanitation and hygiene in achieving health and productivity benefits. Interventions are more successful in

> **Box 1.4 Institutional responsibilities under the WASH MoU**
>
> **Ministry of Water and Energy (MoWE)**
>
> Provision of safe and adequate drinking water for human consumption and domestic use, from source to distribution for communities, schools, and other institutions; water quality monitoring; training of WASHCOs, teachers, parent–teacher associations (PTAs) and others; and operation of the National WASH Coordination Office.
>
> **Ministry of Health (MoH)**
>
> Provision of sanitation facilities in schools and institutions; introduction of appropriate on-site sanitation technologies; monitoring water quality for consumption before and after scheme commissioning; support and supervision of Regional Health Bureaux.
>
> **Ministry of Education (MoE)**
>
> Ensuring water and sanitation provision in schools; supporting establishment of WASH clubs in schools; incorporation of WASH in school curriculum and activities; facilitation of WASH training for teachers and PTAs; and mobilization of school communities to promote hygiene and sanitation in their communities alongside the health sector.
>
> **Ministry of Finance and Economic Development (MoFED)**
>
> Communication with WASH sector ministries on WASH funding programmes and provision of updates on fund disbursement and settlement; ensuring funding transfers to regions are based on action plans approved by the National WASH Steering Committee; ensuring timely programme fund disbursement and settlement; and ensuring financial reporting from *woredas* and regions disaggregated for water supply, hygiene, and sanitation.
>
> **UN agencies, financing institutions, CSOs, and other WASH Development Assistance Groups (DAGs)**
>
> Assignment of representatives to Steering and Technical Committees; facilitation of enabling environment to enhance programme implementation; participation in, and assistance in organization of, the WASH Multi-stakeholder Forum, Joint Sector Review and other WASH fora; and support to the WASH sector in raising financial, technical, and material support to meet MDG and GTP targets.
>
> *Source:* adapted from MoWE et al., n.d.

reducing morbidity and mortality, and also more cost-effective, when coordinated (MoWR, 1999).

Action started in 2005 following the European Union (EU) country dialogue, when integration of water and sanitation was promoted as part of an emerging sector-wide approach (SWAp) agenda and an MoU was signed (revised in 2011, see above). National WASH steering and technical committees were formed from relevant departments of the three ministries, with representation from donor groups and NGOs. Their mandate was to approve budgets, review progress, and select *woredas* that qualify for direct support from the pooled fund. A national WASH coordination unit was also established to oversee integrated planning and reporting of WASH, budget management, and M&E. This is housed in the MoWE and staffed by representatives of the three ministries. Joint annual and bi-annual sector review processes, including the multi-stakeholder forum (MSF) and joint technical review (JTR) missions, were also initiated. At lower levels, regions were also required to develop structures to integrate planning, implementation, monitoring and reporting, and *woreda* WASH teams and community-level WASHCOs were established.

WASH integration objectives have been partially achieved. The MSF provides a platform for sharing experience, wider stakeholder consultation, and policy priority setting, while JTRs have helped to identify and scale up successful WASH service-delivery approaches. A sector-wide M&E framework and the NWI have been initiated as a result of joint-sector review and consultation processes and are recognized as significant developments (Chaka et al., 2011). However, integration has been achieved primarily at national level, rather than at the implementation level where it is most needed (AMCOW, 2011).

Harmonization

As noted above, the new WASH approach aims to move away from discrete WASH projects towards a fully programmatic approach, ideally leading to one WASH plan, budget, and report. The WIF acknowledges that this remains an aspiration, but progress is being made with government and donors agreeing, in principle, to work towards a single, shared system for planning, budgeting, financial management, procurement, information management, and M&E. There are also plans to establish one WASH account – the CWA – to include all donor and government WASH contributions.

Alignment

Alignment has two dimensions. First, donors should base their support on the partner country's development priorities, policies, and strategies ('policy alignment'). Second, aid should be delivered through country systems, rather than parallel structures ('systems alignment').

The WIF reflects this in that major donors and the government agree that the national WASH programme will be aligned with the priorities, policies, and strategies of the respective ministries as outlined in relevant sector development plans. The programme will also use the Ethiopian Government's administrative systems, standards and procedures for financial management and procurement, both vertically (from community through to federal level) and horizontally (across different WASH sub-sectors).

Some donors, notably the World Bank and DFID, now channel funds through the MoFED using a single programme implementation and financial window (Channel 1b, see Box 1.2). However, many donors still use parallel structures, creating high transaction costs and stretching the limited capacity of the MoWE and regional bureaux. A joint aid budget review in 2008 showed that 86 per cent of the treasury budget and 80 per cent of the food security programme budget managed by government were used, but only 48 per cent of the foreign grant budget had been spent. This budget, also managed by government, suffered disbursement delays, difficulties in obtaining no-objection approvals for procurement operations, and cumbersome donor financing conditions (MoWR, 2008). Despite evidence that funds delivered through government systems are used more fully, donors remain reluctant to align completely. Reasons include lack of confidence in national financial management systems, the politics surrounding budget support, and vested interests in the status quo among both donors and government (AMCOW, 2011).

Partnership

National water sector policy highlights the need for partnership across the public and private sectors, donor agencies, NGOs, and communities in the delivery of WASH services. In addition to government-led WASH coordination, donors and government coordinate through the DAG, while NGOs come together under the Consortium of Christian Relief and Development Associations (CCRDA) water and sanitation forum (CCRDA, 2010). Project-based and multi-sectoral coordination structures such as the Millennium Water Alliance and the Ethiopian WASH Movement have also been created for joint advocacy and learning across government and non-government actors.

Current policy emphasizes the role of the private sector in supply chains, in the operation and maintenance of schemes, and in the management of urban WASH services. However, the private sector in Ethiopia remains underdeveloped. The policy environment has not encouraged entrepreneurs, and physical distances, limited infrastructure, general poverty, and low purchasing power of rural inhabitants limit profits and increase risks for businesses.

Private-sector organizations also face difficulties accessing bank loans because of high collateral requirements, while government tendering processes are bureaucratic and demanding. Small artisans at *woreda* or *kebele* (community) level are often unable to access loans as there are no finance institutions in their areas, and they struggle to meet requirements for formal registration.

Decentralization and the mainstreaming of Community Development Funds

Decentralization is a fundamental principle of Ethiopian policy in the WASH sector and beyond. The argument is that decision-making needs to occur with or close to end users to ensure that services fully meet local needs and are sustainable.

Autonomous local institutions have been created to manage water resources and provide water supply services, including utility agencies in large towns, private sector and local artisan associations, water user associations (WUAs), and water committees at scheme level. A key problem is that they typically suffer from limited human capacity, skills, and knowledge, both in technical and managerial terms, and do not always provide the services for which they were established (Arsano et al., 2010).

As part of the decentralization process, *woredas* became the lowest level of government, responsible for decisions on public spending and the provision of services. Regional water bureaux formulate policy appropriate to their own development and play a leading role in managing development interventions. *Woreda* water desks are in charge of investment planning, monitoring, and technical assistance to service providers, at least for lower-end technologies. *Woredas* receive block grants from the central government and decide how to use these grants within broad criteria set by the MoFED. In rural areas, WASHCOs or WUAs operate water systems and promote sanitation, supported by *woreda* and regional water and sanitation government staff (Calow et al., forthcoming).

The Community Development Fund (CDF) approach takes decentralization of WASH services a step further by devolving responsibility for scheme development, construction and management to WASHCOs, including management of capital funds and procurement. Following strong performance in Amhara and Benishangul-Gumuz, with functionality rates of 94 per cent (Chaka et al., 2011), the CDF will be mainstreamed under the banner of community-managed projects (CMP) as a primary service-delivery approach, where communities have the necessary capacity. A bilateral donor project will provide a team of experts to support *woreda* water offices in building community capacity and demand for the CDF approach. Where community capacities or demand are lacking, the main service-delivery option will remain traditional *woreda*- or regionally-implemented projects.

Four modalities for financing and managing community and institutional (school) WASH projects are outlined in the WIF (FDRE, 2011):

- *CMPs* using the CDF approach outlined above.
- *Woreda-managed projects* with WASHCOs or institutional WASH committees involved in planning, design, implementation, and management of schemes. The *woreda* WASH team will be project manager with responsibility for contracting, procurement, quality control, and handover to the community. Larger programmes involving borehole drilling will continue to be carried out by regional government, capitalizing on economies of scale.
- *NGO-managed projects*, varying considerably in approach and scale. Typically NGOs either administer external resources on behalf of the community, following the *woreda*-managed project model, or make external resources available to the community directly or through a

micro-finance institution, for user-group management in an arrangement similar to the CDF.
- *Self-supply*, in which the initiative and investment to build or improve facilities comes from individual households. Self-supply initiatives are 'off budget', but will be documented in the NWI, incorporated into *kebele* and *woreda* WASH plans and reports, and supported with training and technical assistance.

Cost recovery

While not mentioned specifically in the WIF, the National Water Resources Management Policy treats water as both an economic and social good.

In urban areas, tariffs for water supply services are expected to follow full cost recovery principles, including capital costs, operation and maintenance (O&M) costs, depreciation, and debt servicing, with subsidies for poor users. Urban tariffs are tied progressively to consumption levels, with special flat rate tariffs for communal services such as handpumps and public stand posts, used mostly by the urban poor. In reality, however, urban water supply relies on public subsidy.

In rural areas, tariffs are supposed to recover O&M alone, with capital costs borne in full, or in part (depending on scheme type) by government or other financing agencies (MoWR, 2001; FDRE, 2011).However, the assumption that significant revenue can be generated from users and that rural water committees have the capacity to turn this finance into sustainable service delivery is questionable. Funds are often inadequate, technical capacity to maintain schemes often weak, and spare parts supply chains rarely reach lower levels. This means that O&M must be subsidized from public sources, reducing the budget for capital investments (AMCOW, 2011).

A zero subsidy approach is followed for household on-site sanitation, with government resources only going to sanitation and hygiene promotion through the HEP, and to institutional sanitation in schools, health facilities, etc.

Box 1.5 summarizes key principles for service delivery in the updated sector MoU. Some tension may occur, for example on the prioritization of low-cost solutions, while the requirement for assurances of community management capacity may be incompatible with the prioritization of marginalized communities.

Outstanding sector challenges

Strength of *woreda* water offices and WASHCOs

Achieving sustainable services under the four approaches set out in the WIF requires that WASHCOs have the capacity to select appropriate technologies, participate in scheme planning, collect and manage user payments, and carry

> **Box 1.5 Key principles in WASH implementation**
>
> - Adhere to Water Resources Management Policy and Strategy and the GTP (2011–15).
> - Ensure integration of WASH activities at community, *woreda*, regional, and federal levels.
> - All WASH activities to comply with the revised WIF.
> - WASH implementation should respond to demand.
> - Cost recovery should be used to assure a sense of ownership and cost sharing used to fill the financial gap in order to achieve the UAP. Priority should be given to communities that front-load their share of financial responsibilities.
> - Feasible and simple schemes should be prioritized.
> - Schemes should be designed to last a minimum of five years.
> - In rural areas, no scheme should be built without assurance that user communities have the required skills and financial systems to collect and manage user contributions for O&M.
> - An assessment of the local socio-economic situation is central to any WASH intervention.
> - Water is key for marginalized communities and their needs should be prioritized, despite challenges in finding technical solutions and reliable resources.
> - Ultimately, all communities and institutions will be supplied with water, but implementation cannot happen everywhere at the same time. The most cost-effective activities will be prioritized, including prioritizing the expansion of existing schemes or rehabilitation of non-functional schemes over building new schemes.
> - Stakeholder roles and responsibilities across all implementation phases should be clearly defined in advance.
> - Ultimately, it is the communities' responsibility to sustain schemes. Communities need information about feasible technical options and cost implications, both for construction and O&M, to make informed choices.
>
> *Source:* MoWE et al., n.d.

out maintenance and repairs. This places a significant demand on *woredas* to support WASHCOs and build their capacity in areas ranging from scheme design to procurement and financial management, while implementing schemes, carrying out repairs, and monitoring services. The effective integration of water service delivery with sanitation and hygiene promotion at local level also requires institutional strengthening.

Significant investment is needed in capacity building and strengthening of technical expertise at *woreda* level to achieve and sustain sector targets. Human resource shortages are particularly acute in newly-emerging regions (e.g. Afar), and in remote rural *woredas* (Arsano et al., 2010).

Technical and vocational training centres (TVETCs) were opened to train *woreda* water technicians following decentralization in 2004 (MoWR, 1998; MoWR, 2001; MoWR, 2002). However, the training is mainly classroom-based, and most *woredas* still have only a few, often junior, staff with little practical experience. Motivation at *woreda* level tends to be low due to low pay, weak line management, limited capacity-building support, and low budgets.

To address these weaknesses, the WASH programme has developed a capacity-building strategy to:

- increase skills, knowledge, attitudes, and confidence of individuals from water users at community level, to federal managers, private-sector stakeholders, and members of diverse committees;
- support and strengthen the institutional development of new WASH structures at different levels;
- support the capacity of operational systems to achieve harmonized planning, financial management, procurement, reporting, and M&E systems;
- strengthen teamwork, communication, and collaboration across sectors and among all stakeholders;
- strengthen supply and logistical support by supporting private-sector actors to enhance scheme sustainability;
- strengthen strategic sector support to inform WASH policy, implementation, and coordination through strategic studies, sector reviews, and support to networks and fora (FDRE, 2011).

To develop greater capacity, the MoWE (2010) has allocated almost $37 million under the revised UAP to build the capacity of its staff, regional Bureau of Water and Energy (BoWEs), water TVETCs, and the Ethiopian Water Technology Centre (which focuses on groundwater). Trained personnel are expected to provide on-the-job training to other staff.

Conclusion

The challenges to Ethiopia's water sector will not be addressed overnight. However, the WASH integration agenda and NWI, backed by government, major donors, and civil society, promises more efficient and effective use of WASH resources so that access to services continues to rise. The emerging focus on rehabilitation and repair of non-functional schemes, supported by NWI data and balancing a historical emphasis on new construction, is also welcome. Despite efforts to ensure the cost-effectiveness of investments and the focus on low-cost technologies, further increases in sector finance are needed to meet the ambitious UAP target, particularly for sanitation.

With national structures in place, priorities need to shift towards local capacity building. Local decision-makers need the autonomy to determine local strategies, backed by appropriate support and resources. This will allow context-specific solutions, provide space for innovation, and enable locally appropriate balances that focus on self-supply and self-reliance as low-cost options, while ensuring that the poor and marginalized are guaranteed access to reliable and safe water supply. Downward accountability to citizens must also be enhanced if decentralization is to deliver more responsive services, and information systems at local level need to be strengthened significantly.

The benefits of a nationally-harmonized WASH approach include efficiency and unity of approach towards a common goal. Practical arrangements to deliver integrated services are needed at local level, including civil society and private-sector partners, but must avoid excessive bureaucracy and allow local flexibility around service-delivery approaches, so that local decision-makers can learn from, and apply, what works best.

References

Abebe, H. and Deneke, H. (2008) 'The sustainability of water supply schemes: a case study in Alaba Special Woreda', *RiPPLE Working Paper 5*, Research-inspired Policy and Practice Learning in Ethiopia and the Nile Region (RiPPLE), Addis Ababa. All RiPPLE papers available from: <www.rippleethiopia.org/> [accessed July 2012].

African Ministerial Conference on Water (AMCOW) (2011) 'Water supply and sanitation in Ethiopia: turning finance into services for 2015 and beyond', *An AMCOW Country Status Overview*, Water and Sanitation Programme (WSP), Nairobi.

Alemu, G., Deme, M., Habtu, S., and Abdulahi, M. (2010) 'Assessment of budget in the water sector: a case study of two selected woredas in Oromia Regional State (Babile and Goro-Gutu woredas)', *RiPPLE Working Paper 19*, RiPPLE, Addis Ababa.

Arsano, Y., Mekonnen, E., Gudisa, D., Achiso, D., O'Meally, S., Calow, R., and Ludi, E. (2010) *Governance and Drivers of Change in Ethiopia's Water Supply Sector*, Organisation for Social Science Research in Eastern and Southern Africa (OSSREA), Addis Ababa, and Overseas Development Institute (ODI), London.

Awulachew, S. B., Yilma, A. D., Loulseged, M., Loiskandl, W., Ayana, M., and Alamirew, T. (2007) 'Water resources and irrigation development in Ethiopia', *International Water Management Institute (IWMI) Working Paper 123*, IWMI, Colombo.

Butterworth, J., Terefe, B., Bubamo, D., Mamo, E., Terefe, Y., Jeths, M., Mtisi, S., and Dimtse, D. (2009) 'Improving WASH information for better service delivery in Ethiopia: scoping report of initiatives', *RiPPLE Working Paper 13*, RiPPLE, Addis Ababa.

Calow, R., MacDonald, A., and Cross, P. (2012) 'Corruption in rural water supply in Ethiopia', in J. Plummer (ed.), *Diagnosing Corruption in Ethiopia: Perceptions, realities and the way forward for key sectors*, chapter 4, World Bank, Washington D.C.

Chaka, T., Yirgu, L., Abebe, Z., and Butterworth, J. (2011) *Ethiopia: Lessons for Rural Water Supply; Assessing Progress towards Sustainable Service Delivery*, IRC International Water and Sanitation Centre, The Hague.

Comprehensive Assessment of Water Management in Agriculture (CAWMA) (2007) *Water for Food, Water for Life: A Comprehensive Assessment of Water Management in Agriculture*, Earthscan, London and IWMI, Colombo.

Consortium of Christian Relief Development Associations (CCRDA) (2010) *Annual Joint CSO Report on Water, Sanitation and Hygiene (WASH)*, CCRDA, Addis Ababa.

Coulter, L., Kebede, S., and Zeleke, B. (2010) 'Water economy baseline report: water and livelihoods in a highland to lowland transect in eastern Ethiopia', *RiPPLE Working Paper 16*, RiPPLE, Addis Ababa.

Deneke, I., and Abebe, H. (2008) 'The sustainability of water supply schemes: a case study in Mirab Abaya woreda', *RiPPLE Working Paper 4*, RiPPLE, Addis Ababa.

Falkenmark, M. (1989) 'The massive water scarcity threatening Africa – why isn't it being addressed', *Ambio* 18(2): 1112–18 <www.jstor.org/stable/4313541> [accessed July 2012].

Federal Democratic Republic of Ethiopia (FDRE) (2011) *The WASH Implementation Framework (WIF) – Summary*, version: 27 July 2011, FDRE, Addis Ababa.

Hutton, G. and Haller, L. (2004) *Evaluation of the Costs and Benefits of Water and Sanitation Improvements at the Global Level*, World Health Organization (WHO), Geneva.

Mekonnen, W. T., Terefe, B., Yimam, T. M., Getahun, A.H., Newborne, P., and Smet, J. (2008) 'Promoting sanitation and hygiene to rural households: the experience of the Southern Nations region, Ethiopia', *RiPPLE Synthesis Paper*, RiPPLE, Addis Ababa.

Moriarty, P., Jeths, M., Abebe, H., and Deneke, I. (2009) 'Reaching Universal Access: Ethiopia's Universal Access Plan in the Southern Nations, Nationalities, and People's Region (SNNPR)', *RiPPLE Synthesis Paper*, RiPPLE, Addis Ababa.

Ministry of Finance and Economic Development (MoFED) (2010) *Growth and Transformation Plan 2010/11–2014/15*, Volume 1: Main Text, FDRE, Addis Ababa.

Ministry of Health (MoH) (2010) *Health Sector Development Programme IV, 2010/1 –2014/15, Final Draft*, FDRE, Addis Ababa.

MoH (2011) *National Hygiene and Sanitation Strategic Action Plan for Ethiopia 2011–2015, Final Draft*, FDRE, Addis Ababa.

Ministry of Water and Energy (MoWE) (2010) *Revision of UAP (Water), Draft Report*, FDRE, Addis Ababa.

MoWE, MoH, MoE, and MoFED (no date) *Memorandum of Understanding signed by the Ministry of Water and Energy, Ministry of Health, Ministry of Education, and Ministry of Finance and Economic Development on Integrated Implementation of Water Supply, Hygiene and Sanitation Program in Ethiopia, Final Revised WASH MoU draft*, FDRE, Addis Ababa.

Ministry of Water Resources (MoWR) (1999) *Ethiopian Water Resources Management Policy*, FDRE, Addis Ababa.

MoWR (2001) *Ethiopian Water Sector Strategy*, FDRE, Addis Ababa.

MoWR (2002) *Water Sector Development Programme 2002–2016*, FDRE, Addis Ababa.

MoWR, MoH, MoE (2005) *Memorandum of Understanding between Ministry of Water Resources, Ministry of Health, and Ministry of Education on the Implementation Modality for Integrated Water Supply, Sanitation and Hygiene Education (WASH) Programs in Ethiopia*, FDRE, Addis Ababa.

MoWR (2008) *Joint Budget Aid Review: Review of Financial and Performance Data for the Water and Sanitation Sector*, FDRE, Addis Ababa.

Nuru, F. (2012) 'Managing water for inclusive and sustainable growth in Ethiopia: key challenges and priorities', *Background Paper for the European Report on Development 2011/12*, ODI in partnership with the Deutsches Institut für Entwicklungspolitik (DIE) and the European Centre for Development Policy Management (ECDPM).

Slaymaker, T., Adank, M., Boelee, E., Hagos, F., Nicol, A., Tafesse, T., Tolossa, D., and Tucker, J. (2007) *Water, Livelihoods and Growth: Concept Paper*, RiPPLE, Addis Ababa.

Tucker, J., Calow, R., and MacDonald, A. (forthcoming) *Exploring Drivers of Water Use and their Policy Implications: Quantitative Findings from a Highland to Lowland Transect in Ethiopia*, unpublished manuscript.

World Health Organization (WHO) and United Nations Children's Fund (UNICEF) (2012) *Progress on Sanitation and Drinking Water: 2012 Update*, WHO Press, Geneva.

World Bank (2006) *Ethiopia: Managing Water Resources to Maximize Sustainable Growth: A World Bank Water Resources Assistance Strategy for Ethiopia*, World Bank, Washington D.C.

World Bank (2009) *Ethiopia Public Finance Review*, World Bank, Washington, D.C.

About the authors

Eva Ludi is a Research Fellow at the ODI in London. She has over 20 years of experience in policy-oriented and applied research and advisory services on natural resource governance, with a special focus on: water, soil and protected areas; climate change, water and food security linkages; climate change adaptation, adaptive capacity and livelihoods in complex rural environments; and livelihoods-focused sustainable development. She was Research Director for RiPPLE from 2009 to 2011.

Bethel Terefe is a gender and development specialist with a background in political science and regional and local development. She worked as a Policy Officer in the RiPPLE programme, communicating RiPPLE's research outputs to a policy audience and conducting background policy research. She is experienced in supporting NGO policy engagement activities and facilitating CSO fora.

Roger Calow is Head of the Water Policy Programme at the Overseas Development Institute (ODI) and an Honorary Research Associate at the British Geological Survey. He has over 20 years' experience leading international research and development projects in Asia, Africa, and the Middle East, including two years as Director of the RiPPLE programme in Ethiopia. He leads an interdisciplinary team of eight staff working on water supply and sanitation, climate change and water security, water resources management, and the political economy of sector reform.

Gulilat Birhane is a water resources specialist with several years of experience working in policy analysis, sector planning, and socio-economic research. He has engaged in a wide range of water supply, irrigation, and river basin-development projects in Ethiopia, with particular expertise in rural and urban water supply financing, institutional arrangements, and capacity building. Gulilat currently works for WaterAid Ethiopia as Director of Policy Research and Sector Support.

CHAPTER 2
WASH sector monitoring

John Butterworth, Katharina Welle[1], Kristof Bostoen and Florian Schaefer

The Millennium Development Goals (MDGs) and the Paris Declaration have emphasized country 'ownership' of aid programmes, leading to more sophisticated monitoring of results – both outputs and outcomes – to justify donor funding decisions and promote accountability. This chapter examines the monitoring experiences of the water, sanitation, and hygiene (WASH) sector in Ethiopia from the perspectives of global, national, and local actors. Focusing on rural water supply, it reviews the advantages and disadvantages of the key monitoring efforts, including the World Health Organization (WHO)/United Nations Children's Fund (UNICEF) Joint Monitoring Programme (JMP), using national data collected through household surveys, regional inventories, updates used by the government to prepare official sector reports, and the Ethiopian National WASH Inventory (NWI). The chapter unpacks these approaches to examine why the different methods generate different numbers for use, access or coverage of rural water supplies. The chapter argues that the variations between these estimates will probably persist. It also argues for critical analysis of the use of different methods, and provides an understanding of the perspectives of organizations generating the results. It concludes that the future is unlikely to be about one all-encompassing monitoring system, but rather different parallel monitoring processes at global, national, and local levels.

Introduction

Three different institutional perspectives are explored in this chapter, focusing on WASH monitoring at the following levels:

- the global level – comparing progress on access to safe water and sanitation across countries;
- the national level – emphasizing national policy-making and demonstrating progress on the facilitation, regulation and monitoring of WASH activities across a country;
- the regional and *woreda* (district) levels where the focus is on service delivery and responding to unmet needs, including people without water and sanitation facilities or those with failed delivery systems.

Insights from these different levels are drawn together to highlight how we can link and learn from the monitoring efforts, and also to demonstrate when they, and the needs they serve, are just too different to be integrated.

The MDGs have renewed interest in measuring results (Picciotto, 2002; Kusek and Rist, 2004) which, for water and sanitation targets, means obtaining reliable coverage[2] figures. In 1990, the WHO and UNICEF started to collaborate on the JMP which tracks global progress towards the water and sanitation MDGs on a country-by-country basis (WHO and UNICEF, 2010a). As in other countries, some JMP estimates for Ethiopia differ significantly from the official government figures used for planning and policy-making. Why are there discrepancies between the internationally and nationally reported statistics, and what are the consequences? This chapter attempts to address these questions.

Service delivery requires better information, given the ongoing development of new WASH infrastructure and a growing 'sustainability' crisis with more systems breaking down on a regular basis (see Chapter 5 and Moriarty et al., 2009). Regional and local governments need more than just overall baseline information on their WASH infrastructure and assumed levels of coverage. They need to know where there are unmet needs and how they can be addressed, and when, where, and why systems have deteriorated or failed so that resources can be allocated effectively. This chapter explores the different WASH monitoring information required at local level, and the ability of local actors to use that information.

International WASH monitoring and Ethiopia

Global WASH sector monitoring was placed firmly on the international agenda during the International Drinking Water Supply and Sanitation Decade (1981–90), with WHO assuming responsibility for collecting, collating, and publishing global sector information. Data provided by each government through questionnaires were meaningful at country level, but hard to compare at international level. As a result, WHO and UNICEF created the JMP to monitor global progress towards MDG 7 (Target 7c related to water and sanitation was added at a later date).[3] Since 2000, the JMP has based its reporting on user information gathered from household surveys undertaken by national statistical agencies, rather than data for service provision gathered by government ministries.

Contrary to common perceptions, the JMP does not collect data. It relies on existing household surveys that are seen as nationally representative, and that include questions on the types of drinking water and sanitation facilities used by households. This information is used to determine the percentage of households/people using *improved* drinking water sources and *improved* sanitation facilities, where improved requires a certain standard to be met. An improved drinking-water source, for instance, is one that: 'by nature of its construction or through active intervention, is protected from outside contamination' (WHO and UNICEF, 2010a).

WASH SECTOR MONITORING 51

Recent headline figures published by the JMP for Ethiopia on 'use of improved water facilities' are summarized in Table 2.1.[4] These 2010 estimates are extrapolated from surveys conducted over the past two decades (see Figure 2.1), rather than any recent survey (WHO and UNICEF, 2012). Box 2.1 details how JMP figures for Ethiopia have been calculated. These figures have been contested because they show much lower coverage than the government's official statistics. For rural water supply coverage particularly, the two sets of figures have diverged noticeably since 2000 (Figure 2.1).[5]

Table 2.1 Rural and urban water supply coverage for Ethiopia, 2010 (national and international reported figures)

	MoWE water access coverage (%)	JMP use of improved water facilities (%)
Rural	65.8	34
Urban	91.5	97
Total	68.5	44

Sources: Ministry of Water and Energy (MoWE), 2011a; WHO and UNICEF, 2012

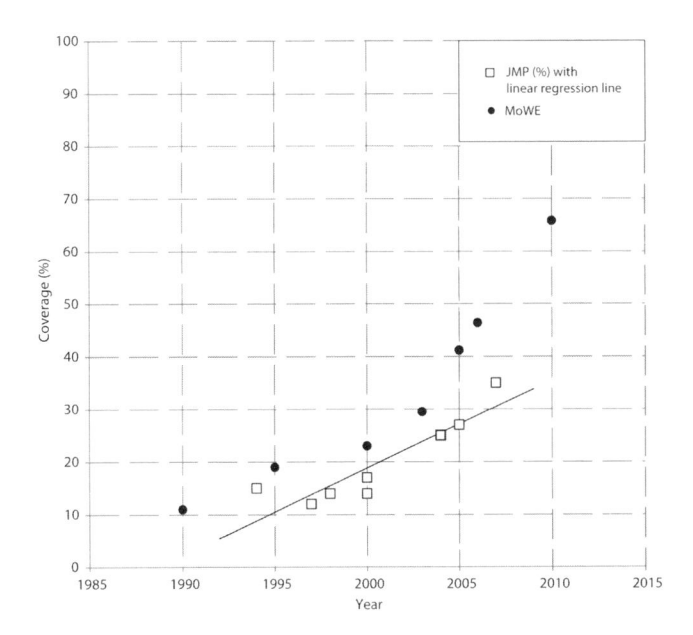

Figure 2.1 Rural water supply coverage (national and international reported figures)

Source: author's own, using data from the JMP (WHO and UNICEF, 2012) and the Ethiopian Government (AMCOW, 2011; MoFED, 2003, 2005, 2007a, 2007b; MoWE, 2011a; MoWR, 2009; MoWR et al., 2006; Rahmato, 1999)

> **Box 2.1 Coverage as defined by the Ministry of Water and Energy**
>
> Ethiopia's Universal Access Plan (UAP) defines the minimum standards for rural areas to be at least 15 litres per capita per day (lpcd) of safe water, available within 1.5 km of the home. However the terms 'access' and 'coverage' are often confused. In the revised UAP (MoWR, 2009), the MoWE defines 'water supply access coverage' as *potential* access while 'water supply coverage' refers to the situation where people *use* 15 litres per capita per day from a source within 1.5 km that meets WHO quality standards. The former definition is the current basis for data collection and reporting, while the government is working towards the latter definition. The method used does not actually monitor the different elements in the national target, i.e. quantity (15 lpcd), quality (potable) or distance (1.5 km in rural areas).
>
> In any given area, whether a *woreda* or entire region, calculations of coverage according to federal guidelines use the number of water systems to calculate the number of people served. They use fixed multiplication factors representing the average number of beneficiaries per scheme type (for example, 270 people for a hand-dug well fitted with a handpump, 457 people for a drilled shallow well fitted with a handpump, and 3,313 people for a deep well with a piped distribution system). Coverage is then calculated by dividing the population that could potentially be served, by the total population estimate for the same area (MoWR, 2009).
>
> This standard methodology is crude but does have its advantages. Counting water sources is relatively simple, reliable, and low cost. An update is derived easily by adding the numbers of people assumed to be covered by new systems as they are constructed. However, there is no means to discount the systems that are broken down. The inclusion of schemes that are non-functional can lead to overestimates, and there are major concerns about the reliability of multiplication factors that ought to be based on extensive, regularly published, and up-to-date research. By definition, standard factors lead to overestimates and underestimates in different places. This method is most unreliable at local level and where systems serve larger numbers of users, such as those with distribution networks (with greater variance around the average assumed number of users). Overestimates are also generated where sources are not distributed in line with the distribution of the population, but are clustered in specific locations (e.g. along roads where access for drilling rigs is better).

How accurate are JMP estimates for Ethiopia?

The JMP does not report information from individual surveys but uses all available data points to draw a trend line as shown in Figure 2.1. Reported estimates are taken from the trend line even if a data point is available for that given year. Values on the trend line are seen as more accurate as they smooth out any errors from individual surveys.

In Ethiopia, the JMP relies upon data collected by the Ethiopian Government's Central Statistical Agency (CSA), such as the United States Agency for International Development (USAID)-funded Demographic and Health Surveys (DHS), the World Bank's Welfare Monitoring Surveys (WMS), and the national census. Some long, bureaucratic delays in accessing these data have caused problems, as having more recent data points would reflect accelerated national efforts to increase coverage. The most recent survey used in deriving Ethiopia's current coverage estimate dates from 2007 (see Table 2.2).

Table 2.2 Data points for rural water supply coverage (national and international reported)

	MoWE[1]	JMP sources[2]	JMP source reference[3]
1990	11		
1994		15	CEN94
1995	19		
1997		12	WMS97
1998		14	WMS98
2000	23	14	DHS00
2000		17	WMS00
2003	29.5		
2004		25	WMS04
2005	41.2	27	DHS05
2006	46.4		
2007		35	CEN07
2010	65.8		

1 Official reported water access coverage (%) for selected years from: Rahmato, 1999; MoH et al., 2006; MoFED, 2007a, 2007b; MoWR (now MoWE), 2009; Tadesse, 2008; World Bank, 2010.
2 Estimated proportion of the population using improved drinking water sources (%)
3 From original data sources: CEN = Census, WMS = Welfare Monitoring Survey, DHS = Demographic and Health Survey

Major amendments were made to the JMP calculations for water and sanitation for Ethiopia at an Addis Ababa workshop organized by the WHO in November 2011. The amendments took into account errors in the classification of protected and unprotected springs in the 2005 JMP dataset, and new data from the 2007 census that had also been made available to the JMP (WHO, 2011). Addressing errors in the 2005 dataset (which provided estimates for 2008) raised estimated total water coverage for 2012 from 38 to 55 per cent of the population (WHO and UNICEF, 2010b), but inclusion of 2007 census data pushed the figure back down to 44 per cent for 2010 because that data point shifted the regression line significantly. This was the total water coverage estimate for 2010, published in the 2012 JMP report. While this reduced the gap between the JMP estimate and official national estimates to some extent, a major discrepancy remains, highlighting the sensitivity of the JMP estimates to the few recent data points that are used when trend lines are drawn. New data anticipated from the 2010 DHS survey will lead to another adjustment of the figures.

National monitoring

Ethiopia's national target is to reach 98 per cent rural water supply coverage by 2015 – more ambitious than the MDG target – reaching another 18 million rural people in the period 2011–15 through almost 94,000 new schemes, and almost 58,600 rehabilitated schemes (MoWE, 2011a). As the Ministry of Water and Energy (MoWE) has started focusing on increasing the number of water points in Ethiopia, it has also developed a way of calculating coverage figures that differs to that used by the JMP. The MoWE estimates coverage by the number of water schemes multiplied by their 'design capacity', meaning the number of people each scheme can serve in theory, regardless of the numbers actually served. In other words, MoWE collects data for, and reports on, service provision (outputs) rather than service use (outcomes) – a methodology summarized in Box 2.1. Official statistics for the water sector[6] are based on information pieced together from occasional inventories undertaken by the regional water bureaux, with newly-constructed schemes added to the figures each year. Results are reported upwards and collated nationally by the MoWE.

Coverage as reported by the MoWE is also shown in Tables 2.1 and 2.2 and Figure 2.1, alongside the JMP coverage figures and data points. In rural areas, in particular, there is a major difference in the two estimates[7] with JMP estimating 34 per cent use of improved sources or about half the MoWE access coverage estimate, namely 65.8 per cent. For the country as a whole, represented in Table 2.1,[8] the JMP estimate of 44 per cent with access to improved water facilities is much lower than the MoWE access coverage figure at 68.5 per cent. With such large differences the MoWE tends to dismiss the JMP access figures (although these are all based on GoE data collected by the CSA) pointing in particular to their failure to reflect recent gains.

Reasons for differences with JMP figures

Differences between the official MoWE and JMP figures are best understood as resulting from differences in definitions, methodological approach, and data access (Butterworth et al., 2010) as well as population estimates and timing which sometimes make it impossible to compare estimates for the same year.

- While the MoWE standard for rural areas is 15 lpcd, the international standard is 20 lpcd. However both approaches lack measurement of the actual water volumes consumed as well as other key parameters to monitor access such as water quality or distance. Crude approximations or proxies are relied on if these are considered at all.
- The JMP has its own specifications for improved sources, including family-owned wells, that must be functional at the time of the survey, while the MoWE counts only communal systems even if they are not functional at the time of the inventory.

- The JMP uses household surveys to assess what facilities people use, while the MoWE monitors the number of schemes and calculates 'water supply access coverage' using assumptions about user numbers per scheme and the estimated population. Both methods have limitations. The number of users of water points rarely matches their design population. Actual data on numbers of users is not widely available in Ethiopia, but similar methodological discussions have shown very significant discrepancies in other places.[9] The sampling strategy used for household surveys seeks statistical representation at the national level, but disaggregation – while possible at sub-national level – does not extend to the local and/or implementation level. These household surveys do not yield information on water sources such as the functionality data needed for operational purposes.
- Ethiopia, like many other developing countries, suffers from a lack of high-quality data, and institutional hurdles may hamper the sharing of the limited data available. This has been the case between the JMP and the CSA,[10] leading to severe delays. While more recent data were available within the CSA, the JMP could only use the 2005 survey as its most recent data point in its 2010 publication (WHO and UNICEF, 2010b). As a result, the JMP estimates have been less accurate and up-to-date than they could have been had all available data been included.

Given the lack of understanding of the differences between monitoring approaches using user- and provider-based data and their respective strengths and weaknesses, it has been impossible to establish consensus between individuals and organizations working at different levels and with different perspectives on which set of figures is more likely to represent reality. However the question of which estimate is more accurate may not be the right question to ask. It would be more useful to compare and triangulate the results from different monitoring approaches, reflecting on why actors choose certain approaches (Welle et al., 2012), and on how well the approaches inform decisions on access to sustainable WASH services. In the next section we turn to one potential solution: the Ethiopian National WASH Inventory (NWI).

The National WASH Inventory

WASH monitoring in Ethiopia is being transformed by the first National WASH Inventory (NWI) which started in 2010 and is a key element of ongoing sector reforms. The 2010/11 NWI was a very resource-intensive exercise, carried out at an estimated cost of Ethiopian birr (ETB) 200 million (c.US$12 million),[11] using more than 65,000 enumerators. An important aspect of the 2010/11 NWI is that it collected both user and provider data through a sector-specific household and water point census.[12] The driving forces for the NWI include the differences observed between the national and international figures, and federal concerns about results from regional inventories that have been held at different times and using diverse methods. One key objective of the NWI is to determine the access figures in a way that will withstand international scrutiny.

Donors have supported the NWI, as they need a credible set of figures to track the use of more integrated activities and pooled funding. Linked to this is an ongoing move towards a more programmatic approach in the sector encapsulated in the WASH Implementation Framework (MoWE, 2011b) which will guide the integration and implementation of all future WASH interventions in Ethiopia (see Chapter 1). Relying on a single plan, budget and report for WASH – the ultimate aim of the reforms – is only feasible if donors have confidence in the government structures managing their funds. This requires a monitoring and evaluation system that is trusted by the Ethiopian Government and donors alike, especially in providing the necessary information to account for funds for the sector. If all donor funds are pooled into a consolidated WASH account as envisaged, it will no longer be possible for donors to track their funds to specific implementation sites or activities, and overall sector achievements will become the major milestones. Finally, the NWI is also intended to deliver a monitoring system for the WASH sector that could improve service delivery through more evidence-based planning and policy-making. This means that the exercise must be relevant to the needs of the *woredas* where many responsibilities for improving WASH service delivery, although not necessarily the means, have been delegated.

In the next section, we outline the implementation of the NWI and its potential to inform policy and practice. The analysis of the NWI process presented here was possible thanks to the establishment of a working partnership between RiPPLE and the Ethiopian Government's NWI project office, which included the organization of two learning seminars, as well as collaboration with the Water and Sanitation Forum of the Consortium of Christian Relief and Development Associations (WSF-CCRDA) and JMP staff.

Reflections on NWI implementation

Originally conceptualized in the 2008 WASH Monitoring and Evaluation (M&E) Framework, the initial design of the NWI included 10 data collection forms and two additional summary formats for household-level information (MoWE, 2011c; see Box 2.2). The formats and choice of questions were influenced strongly by a pilot study in eight *woredas* across Ethiopia, undertaken as part of the introduction of WASH planning within UNICEF-supported areas. Some of the more challenging and technical aspects of the NWI were later dropped in consultation with the Ministry of Health (MoH), the Ministry of Education (MoE), and the CSA to allow the use of non-specialist enumerators.

The way the NWI was conceptualized reflects a sectoral development process managed by the federal government that was geared towards the information needs of both ministries and the donor community in Addis Ababa, rather than building understanding of local services and needs

> **Box 2.2 Data collected in the National WASH Inventory**
>
> There are ten forms used in collecting NWI information, of which five relate to rural WASH:
>
> **Form 1**, Safe Water Supply Inventory for Rural and Small Towns: eight parameters including name of scheme, type of water supply, coordinates, estimated total number of households using a scheme and number within 1.5 km, total yield, functionality, and reasons for non-functionality.
>
> **Form 2**, Health Institutions' WASH Facilities Inventory: eight parameters including type and functionality of water supply facilities, type of latrines, and whether separate facilities exist for men and women.
>
> **Form 3**, Schools' WASH Facilities Inventory: nine parameters, similar to form 2, with the addition of student numbers.
>
> **Form 4**, Inventory of Household Hygiene and Sanitation – Rural and Urban Areas: six parameters including name of household, gender of household head, type of latrine/toilet facility, evidence of use, handwashing facilities, and safe water management in the home.
>
> **Form 5**, Inventory of Household with Source of Drinking Water – Rural and Urban Areas: three parameters including name, gender of household head, and main source of water for household (eight possible responses).
>
> *Source:* MoWE, 2011c

among *woreda*-level water staff. Donors were consulted on the development of the data collection instruments as were line ministries and the CSA, while regional governments had limited opportunity to provide feedback. There was limited consultation at lower levels and this was probably rather too abstract for potential users.

The consequences are clear when looking at the data collection process. Rather than having water sector staff collect data on water supply schemes, the NWI relied mainly on *kebele* (community) staff from other sectors (health extension workers and teachers) as enumerators. While this allowed faster implementation of the NWI by overcoming staff and enumerator shortages, it often deprived *woreda* water officers of the opportunity to visit schemes by themselves. The bias towards national rather than local needs is also reflected in the questions asked under the NWI, with some opportunities missed to generate information relevant for local planning (see Box 2.3).

Ethiopia faces a thorny problem of balancing the need for more data, especially at the local level, with the need to make data collection manageable. A further problem is the limited capacities to use the data that are available. The results of the NWI were still pending at the time of writing, and delays with data entry and analysis, especially for nationwide household level data, are major concerns. It is not yet clear when the new national access figures will

> **Box 2.3 Rural water supply in the National WASH Inventory**
>
> For rural water supply, Form 1 'Safe Water Supply Inventory for Rural and Small Towns', (MoWE, 2011c) captures data on the type of water supply, the Global Positioning System (GPS) coordinates of each scheme, the estimated total number of households using the scheme, the number of those households estimated to live within a 1.5 km radius, and the total yield of the scheme. The format also records scheme functionality and categories of reasons for non-functionality. This is all useful for calculating coverage. However, it does not provide much of the data required for day-to-day operations at the *woreda* level. For example, there is no information on more technical aspects related to scheme failure, on the existence and functioning of WASH committees, or details on their financial management. Such detailed technical and management information may be irrelevant for monitoring at the national level but is vital for *woreda* water supply officers. In addition, the NWI does not give each water scheme a unique reference number, but refers to them by GPS coordinates. Without a unique reference system, it will be more difficult to use and update scheme information and combine different data sources.

be published using this improved survey, or when data will be available for use at lower levels. The way the NWI was conceptualized and implemented raises questions, therefore, about whether results will be available in time to inform implementation of WASH services. The potential for future use of data at regional and local level is the subject of the next section.

Regional and local data collection and use

Many WASH responsibilities in Ethiopia are decentralized to region and *woreda* levels as a result of the constitutional emphasis on federalism and devolution. From the regional perspective, rural water supply data are required to inform political decision-making, to guide budget allocations to *woredas*, and to improve scheme functionality through better maintenance. Major repairs are often organized by regional government, given the greater capacities and equipment available at that level. Meanwhile, the *woreda* is at the frontline of service delivery, with responsibility for managing and maintaining existing water schemes and establishing new ones. However, the ability, capacity, and resources available vary strongly between *woredas*.

Although the NWI is a federal initiative, its achievement requires serious effort by regional governments and administrations, with regional government staff coordinating data collection in collaboration with zone, *woreda*, and *kebele* staff. In addition, regional governments have covered part of the costs with support from regional NGO partners, including from the Research-inspired Policy and Practice Learning in Ethiopia and the Nile Region (RiPPLE) programme. These costs have covered the provision of in-kind assistance in the form of additional staff for supervision, vehicles, and other logistics. Data entry and analysis are also focused on the regional level, creating a significant burden, with the entry of household-level data expected to take over two years in some regions, making it essential to prioritize the scheme-level data needed for coverage calculations.[13]

Looking back at regional WASH inventories

Before the current NWI initiative, regional inventories were undertaken by some regional governments, albeit in an irregular and ad-hoc fashion (Etherington et al., 2008). Executed independently from the federal MoWE, these also focused on estimating coverage but sometimes included other indicators. One example is the 2009 *Woreda* Inventory Survey undertaken in Southern Nations, Nationalities, and People's Region (SNNPR). Disagreements over coverage figures among *woreda* and regional-level sector staff over the previous data had generated political interest in SNNPR and led to the commissioning of a fresh data collection exercise by the regional cabinet. Discrepancies between the reasonably high reported rates of coverage and large numbers of un-served people (that needed tankered supplies) were brought into sharp focus during a severe water shortage in 2009.

The results of the 2009 regional SNNPR inventory were analysed in at least three different ways. Coverage calculated using the standard water supply access coverage method generated a rural coverage estimate of 36 per cent. An alternative non-standard method which has been tested but not officially approved, is to elicit the estimated number of users[14] of each source (this resulted in an estimate of 30.9 per cent) and also specifically those living within 1.5 km (resulting in 20.8 per cent coverage). Under this alternative method, numbers of users are estimated by asking the communities involved, instead of using multiplication factors for users of different schemes to derive an access estimate. To be reliable this method requires careful questioning as it assumes that the managers of water systems, e.g. WASH Committee (WASHCO) members, can estimate accurately how many people use a system as their main source of domestic water, and their distance from the source. Although not used officially, these calculations illustrate the sensitivity of the results to the method. The NWI also collects the data needed to use these different methods. The calculation of figures in the 2009 inventory illustrates that access figures, although seemingly objective, are negotiated in reality and are subject to political considerations. From this perspective, an interesting question is: which method will be used to calculate the NWI data?

In the past, many regional inventory data have not been used and are not easily accessible (Box 2.4 discusses a RiPPLE effort to address this problem). One key issue is the underestimation of the resources needed after survey completion, and the investment needed to maintain records and archives. Welle et al. (2012) examine why costly and human-resource intensive baselines tend to be underused. The fear is that the NWI will suffer a similar fate, and this will be determined at the regional and *woreda* levels.

> **Box 2.4 Encouraging local use of data**
>
> RiPPLE experimented with stimulating the local use of currently underused data, building upon the SNNPR Woreda Inventory Survey of 2009. In collaboration with four *woredas* in the region, RiPPLE assessed and strengthened capacities to analyse and use these WASH data in *woreda* Water, Mining and Energy (WME) offices. The assessment showed that *woreda*-level water staff members tend to lack basic analytical and computer skills needed for water access-related calculations, but found strong interest in analysing WASH data. Six days of training spread over several months on WASH indicators, calculation methods, and presentation skills, using mainly Microsoft Excel but also Google Earth-based maps (primarily WaterAid's Waterpoint Mapper software), helped staff to better understand the situation in their *woredas* and plan for future interventions. Working with their own data, *woreda* water experts experimented with different calculation methods. Participants were surprised by the large differences in coverage figures that resulted, with the inclusion – or non-inclusion – of functionality having a major impact.
>
> The sessions showed that, given this training, spreadsheet software like Excel can be used successfully by some – but not all – *woreda*-level staff. One *woreda* WME office had gone on to use the report card produced to lobby their *woreda* council for more funds. This approach should not, however, be seen as an alternative to the new Microsoft Access-based software for the NWI. A database is required to enter and store data reliably and safely. However, spreadsheets make it possible for *woredas* to produce locally-tailored information and plans in a timely manner. How the NWI makes the database software available at *woreda* level will be important to future use of the data. In some *woredas*, updating data was stimulated by the training courses and undertaken on the initiative of participants. Where this happens there should be a mechanism for updated data to be captured and fed upwards.
>
> *Source:* Butterworth and Dimtse, forthcoming

Potential to use NWI data at regional and *woreda* levels

Making data accessible has been touted as a key feature of the NWI. Data are to be stored and made available through a WASH Management Information System that is being developed. A much more simple Microsoft Access-based interim database has already been designed to enable data entry at regional level and the production of *woreda* report cards, displaying basic information on WASH access through pie charts, graphs, and *woreda* profiles. There is a commitment to make data available electronically within the government through WoredaNet, a satellite network that makes information available to clusters of *woredas*. Uncertainties remain about making data accessible to partners outside the government: there are questions around the type of data and whether raw data or aggregated information will and should be made publicly available. Answers to the data access questions could be an incentive for increasing NGO collaboration and donor funding in the NWI. NGOs were mobilized to support the NWI but without a clear commitment at that time from the government to make the data available to support future planning and reporting. Seeing NWI data in use could encourage support for future updating exercises that must draw on all kinds of capacities – including staff and vehicles – from NGOs. Answers on data access will also be important to ensure accountability and allow NGOs and donors to track sector spending and outputs for reporting back to their constituencies in donor countries.

Using data: needs and capacity at *woreda* level

When comparing the current stage of NWI implementation with the aspirations for decentralized multi-sectoral planning for WASH, a gap emerges between aspirations and available capacity to develop such integrated, *woreda*- or *kebele*-level plans. One important way to use NWI data and resulting information is in the development of local WASH plans. While data are entered into the NWI database at the regional level, *woreda* water, health, and education officers will have to familiarize themselves with the information generated and find ways to work with the data in order to use it as a planning tool. Although the level at which data will be aggregated is not yet clear, it seems crucial to make the NWI data available either as raw data or, if aggregated, at the lowest possible level (*kebele*), to be of most use for developing relevant WASH plans.

For those *woredas* with little exposure to such systems, or IT systems in general, this could be quite a challenge, requiring new sets of skills for the staff involved. In addition to logistical issues, such as how to organize training, and where the database would be hosted, other bottlenecks may emerge with consequences for integrated planning at *woreda* level. If, for example, regional offices enter WASH data into the database, to what extent will *woreda* officers be able to update the data? As mentioned in Box 2.3, one practical issue is that water schemes are not allocated a unique reference number in the NWI.

Participants at a RiPPLE workshop in May 2011 that discussed these issues highlighted the immense differences between individual *woredas* in terms of available staff capacity and logistics (Welle and Bostoen, 2011). Those assisted by donor programmes have benefited from several years of intensive capacity building and logistical support in planning and implementation, including the provision of vehicles, computers, and other basic equipment. *Woredas* without donor assistance, however, have not received any capacity-building support, and sometimes have had no transport or other hardware for their work. *Woredas* that have not yet benefited from such support were found to be in a poor position to use NWI data to undertake formal WASH planning and could benefit from more targeted assistance or inter-*woreda* exchange.

There are also different perceptions of the ownership of NWI data. While the Directorate of Water Supply and Sanitation in the MoWE feels strongly that the *woreda* is the ultimate owner of the current data, the way in which data are currently collected and analysed and the way they will be made available makes *woreda* representatives feel that the data are, in reality, owned at the national level. These different perceptions matter for the future reliability of the NWI because, as discussed above, the current thinking is to develop an IT structure in which the *woreda* will be responsible for the collection, updating, and analysis of the WASH data at local level. It is vital, therefore, that the *woreda* uses, and feels responsible for, local data.

The *woreda* water officers who participated in the workshop had actually collected water supply data for the NWI themselves because of a lack of *kebele*-level staff in their *woredas*. They found the data collection process very helpful for increasing their understanding of existing local water supply schemes, and in developing collaborations with health and education colleagues. Visits to other water schemes allowed water officials to conduct informal discussions with users and WASHCOs. These enabled *woreda* water officers to establish or renew contacts, assess the sanitary, operation and maintenance (O&M), and financial management situation at each scheme, and to provide quick feedback and advice. It was clear that the process of collecting data for the MoWE (as it was perceived) at *woreda* level was viewed as a great opportunity to visit water schemes that had not, in some cases, been visited by water officers for months or even years. It also allowed additional data to be collected that were not needed for the NWI but important for the water officers. This informally 'collected' information often related to the functioning of WASHCOs and, in particular, to the financial management of WASHCO funds.

The fact that formal data collection opened up processes for informal discussions around water schemes shows that the process of knowledge creation does not just follow one single, formal path. In the test phase of the NWI, *woreda* water officers were not encouraged to analyse the data they collected or make use of them in any other way. However, many may well have done so informally outside the NWI process, catalysing local knowledge creation.

Conclusion

There is widespread consensus in Ethiopia that better WASH sector monitoring is a vital step to improving sector performance. As a result, monitoring is high up on the sector agenda. However, it is not certain that the NWI and the evolving WASH monitoring system will, in practice, provide the conditions for improving sector performance.

One point emerging from WASH monitoring efforts in Ethiopia is that the future will probably not be about one all-encompassing monitoring system, but rather different parallel monitoring processes at global, national, and local levels. The JMP mandate for global monitoring is likely to extend beyond the MDG deadline of 2015, while Ethiopia will always be responsible for its own national sector monitoring. It is likely, therefore, that both these systems will co-exist for the foreseeable future, requiring navigation of the interface between international and national monitoring. Given their fundamentally different methodologies, the two approaches can generate diverging results, as we have seen in the example of rural water supply.

Efforts have been made to better understand the different monitoring processes and viewpoints, and to document, reconcile, and harmonize methods through nationally-focused reconciliation workshops between the JMP and national agencies. However, the differences between these estimates will probably persist and such efforts need to continue. Critical analysis on the use of different methods, and a better understanding of the perspectives of the organizations that generate the results should be helpful.

The NWI and the post-2015 revisions to the global monitoring regime and modified indicators are expected to alter the WASH monitoring landscape in Ethiopia. The development of the NWI is a huge step forward as it includes all relevant national institutions such as the CSA and all relevant line ministries, at least at the national level. The NWI uses data collection methods that have been scrutinized by the CSA and has the potential to generate user-based as well as provider-based data that might be acceptable for inclusion in JMP estimates.

At the time of writing, results of the NWI were not yet published and data entry and analysis of the household level data is expected to take considerable effort yet. Once the analysis is completed and the results are published, the NWI may establish a clear sector baseline using methods that are acceptable to the Ethiopian Government, represented by the MoWE, the CSA, and the JMP, and requiring further collaboration between these bodies at the analysis stage. However, the results and methodologies may also be contested by various stakeholders, including potentially by the regions and donors, and the results might differ significantly from current estimates of access coverage. The interest should then turn to the next steps and the use or non-use of data.

While a large amount of resources have been consumed in undertaking the NWI, it is not clear how the vast amount of data will be updated sustainably in the years to come. It is also not clear how the capacity will be built, particularly at *woreda* level, to actually use the data for relevant WASH planning and implementation. While there might be partial technical solutions to the perennial under-resourcing of data entry and analysis, such as improved data collection tools on portable smart phones that could reduce the data entry burden and associated delays, the governance issues outlined in this chapter are likely to be critical. Updating will be encouraged by active use of the data, but current data is already underused. Given past inability to make full use of existing data, and a danger of politicization as results are generated, the NWI may not prove as useful as it could be at local level without far more support to develop capacities, and clarification of the outstanding questions on aggregation and access.

It is a huge achievement that Ethiopia has managed to mobilize the vast resources required to collect the information for its first NWI. Dialogue between the global and national monitoring levels is improving and the NWI has encouraged better collaboration between the CSA and the line ministries.

The next step is to convince and support the WASH sector to use the available data in the NWI and from other sources. Here the challenge is local. It is about putting available data into practical use to improve actual operations for service delivery, as much as overcoming capacity constraints to make good use of data. That will probably require a greater and more sustained effort than that already expended on NWI data collection.

The NWI provides only part of the data needed for operational management of rural water supplies, and supplementary data will need to be added and combined. The real challenges lie ahead. To meet them we must not only acknowledge the different data needs at different levels, but also engage actors proactively at all levels in the process to use available data for planning and implementation, and to design, implement, and update monitoring tools.

Notes

1. This chapter is based partly on a related article published in the *Institute of Development Studies (IDS) Bulletin* (Welle et al., 2012), and benefits from inputs from Katharina Welle as part of her PhD research on WASH sector monitoring in Ethiopia.
2. Coverage (expressed as a percentage) is used by the JMP as shorthand for the numbers of people using different types of improved water and sanitation facilities. Coverage and access are also used to refer to the official Ethiopian Government estimates of the people with access to a water supply scheme (see Box 2.1 for further discussion). Readers should be aware that the term 'coverage', is measured in different ways by different agencies.
3. Where the targets are to reduce by half by 2015, the proportion of people without sustainable access to safe drinking water and basic sanitation.
4. The Joint Monitoring Programme website at <www.wssinfo.org> provides extensive information on history, methodology and access to data files for each country.
5. This chapter focuses on rural water supply, but Ethiopia's sanitation figures have also been contested. The disagreement centres on different definitions of 'improved sanitation', with the JMP's international standard stricter than that preferred by the GoE.
6. There is no WASH sector as such but rather some cross-ministerial collaboration involving health, education, and finance. Sectoral reforms are underway to instigate a sector-wide approach to programming.
7. The difference is relatively small for urban areas.
8. Results are weighted heavily towards rural areas in accordance with the distribution of the Ethiopian population (82 per cent rural in 2010 according to the World Bank, 2012).
9. In a workshop in Mozambique organized by Unicef with the national statistical office (INE) and the water department (DNA) calculations were reported showing that in rural areas the actual number of users per water point was on average 253 in rural areas and 2386 in urban areas, which in both cases was significantly different from the 500 users per water point used as a design criterion (Bostoen, 2008).
10. CSA has a policy to charge for data, and UN agencies have a policy not to pay.
11. ETB 16.8 to US$1 (1 January 2011 rate).

12 The household-level survey of some 15 million households is unlikely to be repeated because of the associated effort and cost, with future data collection focusing on updating the water point data every two years (personal communication from Tamene Hailu, 2012).
13 One could also ask why household-level data was collected separately and not through the normal census or sample survey, which would have made data analysis more efficient. The MoWE propose not to conduct surveys in this way again, and future surveys might also reduce the data entry burden by deploying new technology available for using smartphones as data entry devices.
14 User numbers were estimated by communal WASH Committees (WASHCOs), who are responsible for the day-to-day management of a water scheme on behalf of the user community.

References

African Ministerial Conference on Water (AMCOW) (2011) 'Water supply and sanitation in Ethiopia: turning finance into services for 2015 and beyond', *An AMCOW Country Status Overview*, Water and Sanitation Programme (WSP), Nairobi.

Butterworth, J. and Dimtse, D. (forthcoming) 'From data to information and knowledge: training at woreda level to promote local use of WASH data', *RiPPLE Working Paper*, RiPPLE, Addis Ababa.

Butterworth, J., Welle, K., Bostoen, K., Chaka, T., Goshu, A. (2010) 'Monitoring WASH in Ethiopia: messages from a sector symposium', *RiPPLE Meeting Report*, Research-inspired Policy and Practice Learning in Ethiopia and the Nile Region (RiPPLE), Addis Ababa. All RiPPLE papers available from: <www.rippleethiopia.org> [accessed July 2012].

Etherington, A., Yemane, Y. and Woldemichael, T. (2008) 'Strengthening of national WASH M&E in Ethiopia', *Draft Mission Report* (unpublished), Ministry of Water and Energy, Addis Ababa.

Kusek, J. Z. and Rist, R. C. (2004) *Ten Steps to a Results-based Monitoring and Evaluation System: A Handbook for Development Practitioners*, World Bank, Washington D.C.

Ministry of Finance and Economic Development (MoFED) *(2003) Sustainable Development and Poverty Reduction Program (SDPRP) Appraisal Report*, Federal Democratic Republic of Ethiopia, Addis Ababa.

MoFED (2005) *Ethiopia: Building on Progress: A Plan for Accelerated and Sustained Development to End Poverty (PASDEP)*, Federal Democratic Republic of Ethiopia, Addis Ababa.

MoFED (2007a) *Ethiopia: Building on Progress: A Plan for Accelerated and Sustained Development to End Poverty. Annual Progress Report 2005/6*, Federal Democratic Republic of Ethiopia (FDRE), Addis Ababa.

MoFED (2007b) *Ethiopia: Building on Progress: A Plan for Accelerated and Sustained Development to End Poverty. Annual Progress Report 2006/7*, FDRE, Addis Ababa.

Moriarty, P., Jeths, M., Abebe, H., Deneke, I., and Dickinson, N. (2009) 'Reaching universal access: Ethiopia's Universal Access Plan in the Southern Nations, Nationalities, and People's Region (SNNPR)', *RiPPLE Synthesis Report*, RiPPLE, Addis Ababa.

Ministry of Water and Energy (MoWE) (2011a) *Revision of UAP (Water) Draft Report*, FDRE, Addis Ababa.

MoWE (2011b) *National WASH Implementation Framework. Draft Report April 2011*, FDRE, Addis Ababa. Available from: <www.mowr.gov.et/index.php?pagenum=10> [accessed July 2012].

MoWE (2011c) *National Water Supply, Sanitation and Hygiene Inventory. Unpublished Formats for Data Collection*, MoWE, FDRE, Addis Ababa.

Ministry of Water Resources (MoWR), Ministry of Health (MoH), et al. (2006) *Final Sector Review Report of Water Supply, Sanitation and Hygiene in Ethiopia*, FDRE for presentation to the 1st Multi Stakeholder Forum, 11–13 October, Addis Ababa.

MoWR (2009) *Review of Rural Water Supply Universal Access Plan Implementation and Reformulation of Plans and Strategies for Accelerated Implementation*, MoWR, FDRE, Addis Ababa.

Picciotto, R. (2002) *Development Cooperation and Performance Evaluation: The Monterrey Challenge*, World Bank, Washington D.C.

Rahmato, D. (1999) 'Water resource development in Ethiopia: issues of sustainability and participation', *FSS Discussion Paper*, Forum for Social Studies, Addis Ababa.

Tadesse, T. (2008) 'A review of Ethiopia's water sector policy, strategy and program', in A. Taye, *Digest of Ethiopia's National Policies, Strategies and Programs*, Forum for Social Studies, Addis Ababa.

Welle, K. and Bostoen, K. (2011) 'What next after completing the first National WASH Inventory? Suggestions from stakeholders at the national, regional and woreda level', *RiPPLE Meeting Report*, RiPPLE, Addis Ababa. Available from: <www.rippleethiopia.org/documents/info/20110608-workshop-report-may-2011> [accessed July 2012].

Welle, K., Schäfer, F., Butterworth, J., and Bostoen, K. (2012) 'Enabling or disabling? Reflections on the Ethiopian National WASH Inventory process', *IDS Bulletin* 43(2): 44–50.

World Health Organization (WHO) (2011) *National Consultative Workshop on Joint Monitoring Programme (JMP) and Global Annual Assessment of Sanitation and Drinking Water (GLAAS) 2010, Update Report for Water Supply and Sanitation Data Re-Conciliation*. Report of workshop held 24–25 November 2011, WHO, Addis Ababa.

WHO and United Nations Children's Fund (UNICEF) (2010a) *WHO/UNICEF Joint Monitoring Programme (JMP) for Water Supply and Sanitation: Introduction* [website] <www.wssinfo.org/definitions-methods/introduction/> [accessed July 2012].

WHO and UNICEF (2010b) *Progress on Sanitation and Drinking-water: 2010 Update*, JMP, WHO, and UNICEF, Geneva.

WHO and UNICEF (2012) *Joint Monitoring Programme for Water Supply and Sanitation, Estimates for the Use of Improved Drinking-Water Sources* <www.wssinfo.org> [accessed July 2012].

World Bank (2010) 'Water supply and sanitation in Ethiopia: turning finance into services for 2015 and beyond', *AMCOW Country Status Overview*, World Bank, Nairobi.

World Bank (2012) *Ethiopia Country Data* [website] <http://data.worldbank.org/country/ethiopia> [accessed 15 February 2012].

About the authors

John Butterworth is a Senior Programme Officer at the IRC International Water and Sanitation Centre where he coordinates IRC's Ethiopia Country Programme. He managed IRC's inputs to the RiPPLE project, and coordinated research activities on behalf of the consortium on sector monitoring and self-supply.

Katharina Welle is currently completing a PhD in STEPS Centre on monitoring access to rural water supply in Ethiopia. Prior to her studies, she worked for the Overseas Development Institute (ODI) and for the Water and Sanitation Programme of the World Bank in Kenya, Yemen, and Ethiopia. Her work focuses on monitoring and evaluation and aid effectiveness and governance questions.

Kristof Bostoen works at the IRC International Water and Sanitation Centre where he coordinates activities in monitoring and learning, and training. Previously he studied and lectured at the London School of Hygiene and Tropical Medicine researching the measurement of access and practices within the WASH sector.

Florian Schaefer is a PhD candidate at the School of Oriental and African Studies (SOAS), University of London, researching the emergence and growth of commercial agriculture in Ethiopia. Prior to this, he was an ODI Fellow working as an economist at the Ministry of Water and Energy in Addis Ababa where he supported the development of the National WASH Inventory.

CHAPTER 3

Innovative approaches for extending access to water services: the potential of multiple-use water services and self-supply

Marieke Adank, John Butterworth, Sally Sutton and Zemede Abebe

Community management is the main model for rural water service provision in Ethiopia, as in the rest of sub-Saharan Africa (SSA). The model places communities centre stage in terms of articulating demand and in the design, implementation and operation of schemes. It is similar across countries, differing only in the details, and was intended to match services to demand and strengthen local ownership. However, community management has often failed to deliver on the promise of sustainable services. This chapter assesses the strengths and weaknesses of the conventional community management model in Ethiopia, and then discusses the potential of two complementary approaches for water service delivery recently promoted under Ethiopia's Universal Access Plan (UAP): multiple-use water services (MUS) and self-supply. Both can enhance the benefits from water services and improve sustainability. In spite of their inclusion in the UAP, however, scalable models have not yet been developed. Piloting is recommended, with a focus on how to overcome institutional and capacity constraints to uptake at local level – some of the same constraints, in fact, which have hindered conventional service delivery models.

Community management and the need for innovative approaches

The model of community management for water service delivery has its roots in the International Drinking Water Supply and Sanitation Decade (1981–1990), when large implementation programmes were supported by donors and NGOs, often by-passing government structures in favour of communities and grassroots organizations. This aimed to prevent dependency on (limited) local government capacity and encourage a 'bottom up' approach to implementation that treated communities as active development partners rather than passive recipients of aid. One key element was 'village-level operation and maintenance' (VLOM), with communities trained in basic operation and maintenance (O&M) tasks, recovering costs for repair from users, and therefore, taking ownership of water points. By the mid-1990s, more emphasis was placed on how implementers could prepare communities to take on these management tasks and respond to communities' demand for services (the demand responsive approach). Demand was generally understood in terms of willingness to contribute to implementation and operational costs, as well as articulation of preferences for different service types and levels, albeit from a very limited domestic supply menu. The aim was to enhance community ownership, and better match demand for services with government, donor, and NGO support, resulting in better targeted services, enhanced cost recovery and, ultimately, services that could be sustained by communities over time.

There has been increasing recognition of the challenges of the community-management approach since the early 2000s (Whittington et al., 2008). These include a widespread lack of technical and managerial capacity at community level, the dependency on voluntarism for scheme management, and a frequent lack of post-construction support provided either through government backstopping or the private sector. With communities growing in size and tariff-base, especially in larger rural settlements and small towns, and with piped networks becoming the norm in these settlements, there is an ongoing trend towards the professionalization of community management in many countries. Post-construction support, however, has not yet been addressed in a systematic way (Lockwood and Smits, 2011).

Community management in Ethiopia: prospects for achieving universal access

In Ethiopia, schemes managed under the traditional community-managed approach are now termed 'Woreda-managed projects' under the Water, Sanitation, and Hygiene (WASH) Implementation Framework (see Chapter 1). Communities generally provide labour during the construction of a well, borehole or spring. Water systems are then 'handed over' to communities and O&M is the responsibility of the users, who establish an elected WASHCO (WASH committee in name, although they focus on water supply in practice) to take charge of operation and minor repairs. In multi-village schemes, water boards are established to oversee more complex management tasks, comprising representatives from the village WASHCOs.

Users pay for the water provided through a water tariff, which is retained by WASHCOs to pay for O&M; it is generally sufficient to cover day-to-day operations and minor maintenance. Payment ranges from monthly flat fees to payment per jerry can of water, with poor families exempted from payment in some communities (Chaka et al., 2011). Support and funding for major repairs usually comes from the *woreda* (district), zone or region.

Ethiopia has an ambitious target for rural water coverage, exceeding that enshrined in the Millennium Development Goals (MDGs) of halving the proportion of people without access to improved drinking water supply. Under its recently revised Universal Access Plan (UAP), Ethiopia is committed to increasing coverage to 98 per cent by the year 2015 (MoWE, 2011a), with coverage defined as access to 15 litres per capita per day (lpcd) within a service radius of 1.5 km.

Rural water coverage in Ethiopia was reported by government as 65.8 per cent in 2010, compared with 15.5 per cent in 1991. A Research-inspired Policy and Practice Learning in Ethiopia and the Nile Region (RiPPLE) study in one region – Southern Nations, Nationalities, and People's Region (SNNPR) – found that if current rates of progress can be maintained, the region will reach 98 per cent coverage between 2016 and 2020 based on these official coverage figures (Moriarty et al., 2009). Taken at face value, and given the ambitious nature of the UAP, this is an encouraging finding and would suggest that the traditional community management approach is working. However, these figures are likely to be overestimates. Current coverage calculations are based solely on the number of water schemes, and assume a certain fixed number of beneficiaries for each scheme type, while *woreda* water offices largely have limited analytical capacity and are not able to verify how many people are in fact accessing the minimum standard of 15 lpcd within 1.5 km (see Chapter 2).

The World Health Organization (WHO) Joint Monitoring Programme (JMP) reports much lower coverage than official figures, based on a different methodology and reporting period (see Chapters 1 and 2), and it is likely that real coverage levels lie somewhere between the two estimates. Recent increases in funding for the water sector, and progress over the last two decades, have given rise to genuine optimism in terms of Ethiopia's ability to extend services rapidly (Moriarty et al., 2009). However, attaining universal access remains a significant challenge, not least because the areas remaining un-served are increasingly those where the costs of service provision are highest, due to remoteness, difficult terrain or challenging hydrogeological conditions (e.g. very deep groundwater requiring expensive drilling), or where inhabitants are mobile pastoralists for whom water provision is less straightforward than for settled communities.

Perhaps the greatest threat to the achievement of universal access, however, is the high proportion of scheme non-functionality, which exposes the weaknesses of current approaches to service delivery. RiPPLE research in two *woredas* in SNNPR found that a high proportion of community-managed systems were not functioning. Of the 70 identified community-managed systems in Mirab Abaya, 30 (43 per cent) were non-functional at the time of the study (November 2007 to February 2008), as were 25 (39 per cent) of the 65 networked on-spot distribution points (Deneke and Abebe, 2008). By 2009 the situation in Mirab Abaya had worsened in spite of the construction of nine new schemes, with 49 per cent non-functional including 72 per cent of boreholes and 56 per cent of hand-dug wells (Abebe et al., 2010). Halaba *Woreda* is served by community-managed deep boreholes with distribution networks to water points. Of the 24 boreholes, 10 (42 per cent) were not functioning, as were 40 (62 per cent) of the 65 networked water points (Abebe and Deneke, 2008).

Breakdown of systems was found to be linked to several factors:

- **WASHCO capacity:** WASHCOs lack the capacity to handle funds, resulting in weak cost recovery. While cost recovery of operation and minor maintenance has been successful in some multi-village schemes (Tekalign, 2001), concerns remain about generating enough cash to cover major repairs and replacements in even the best-managed schemes. WASHCOs also lack essential O&M skills.
- **Institutional arrangements:** WASHCOs still lack legal recognition and so cannot open bank accounts. The resulting ad-hoc arrangements for scheme management, with oversight of funds assigned to, or appropriated by individuals, increase the possibility of misuse of funds. A lack of clarity on underlying legislation and policy, in addition to lack of capacity at community and *woreda* level, also creates ambiguity over the roles of WASHCOs, *kebele* (community) administrators, and *woreda* water offices in tackling system breakdowns.
- **Local government capacity:** *Woredas* lack the capacity to provide the required support to the WASHCOs, in quantity and quality (in terms of experience and received training). In addition, even where sufficient and well-qualified staff are available at *woreda* level, they often lack the transport to visit village schemes.
- **Private-sector development:** WASHCOs struggle to access spare parts because supply chains are poorly developed.

Constructing new schemes will not lead to full coverage unless these issues are addressed. If the trends in the two studied *woredas* hold for the region and the nation as a whole, the implications for Ethiopia's ability to achieve its ambitious UAP goals are worrying. Indeed the main conclusion from a synthesis report on RiPPLE research in SNNPR (Moriarty et al., 2009) was that the critical lack of capacity at WASHCO and *woreda* level is a serious challenge to the community management model of service provision in Ethiopia. Initial data from the first National WASH Inventory (NWI) exercises also suggest that around 35 per cent of schemes may be non-functional for at least a few days in the year (MoWE, 2011a). The Rural Water Supply Network (RWSN) estimated that non-functionality of handpumps in Ethiopia is much higher, at around 65 per cent (RWSN, 2012). The sustainability of rural water supply in Ethiopia, particularly under community management, is discussed in more detail in Chapter 5.

New approaches to extend access

Given these challenges – the increasing cost of delivering services and the poor sustainability of many services, the latter due in significant part to low levels of cost recovery and capacity at local level – recent revisions to the UAP (MoWR, 2009) sought to promote a more pluralistic approach to service provision in rural areas. Multiple-use water services (MUS) and self-supply were promoted alongside traditional approaches, with an overall emphasis on the use of low-cost technologies where possible (see Chapter 1). These approaches will not in themselves address the institutional and capacity constraints outlined above, which may ultimately hinder their success (see Chapter 5), but they are intended to enhance ownership and financial sustainability by capitalising on users' own priorities.

MUS starts from the premise that water is needed and used for many different purposes beyond domestic use, such as livestock production, irrigation/market gardening, and small-scale industrial activities. Under MUS, services are designed explicitly to meet these multiple demands. Communities are expected to be more motivated to maintain systems that address their multiple water needs, while income from productive water uses should improve their ability to pay for upkeep and repairs, thus enhancing scheme sustainability.

In self-supply, individual households (or a group of neighbours) invest in improving their own service (Sutton, 2007), usually from a family well, but perhaps from rainwater harvesting and household water treatment. These households are responsible for O&M and share voluntarily with their neighbours to form a small, privately owned communal supply. Thus government is not responsible for investment, but it must nonetheless perform enabling, support, and regulatory functions to ensure that households can access the necessary technologies, and that water quality standards for drinking water are met and maintained if required. As well as offering a lower-cost route to extend access to water, supporting self-supply is expected to enhance sustainability as individual households tend to be more willing to invest in maintaining their own systems than to contribute to the maintenance of community-managed systems (Sutton, 2004).

Multiple-use water services

The reality of multiple uses of water is not new. People in rural areas use water for domestic activities like drinking, cooking, washing, and cleaning, and for productive activities such as backyard gardening, irrigation, livestock-keeping, processing of agricultural products, and small-scale industrial activities, like beer brewing and brick-making. These multiple uses bring multiple benefits. The domestic use of safe water reduces the risk of diseases related to poor quality water, sanitation, and hygiene, while productive use of water can bring additional income and improved diet, leading to increased food and livelihood security (Moriarty et al., 2004; Van Koppen et al., 2009).

However, the traditional approach to service provision typically fails to accommodate these multiple uses in an integrated way. Different sub-sectors tend to focus on different benefits, with the water supply sector focusing on domestic provision to generate health benefits and the irrigation sub-sector focusing on food security and the economic benefits of irrigated agriculture.

This can often result in a failure to capitalize on the full range of potential benefits of interventions and services – opportunities missed, particularly in relation to widespread food insecurity in Ethiopia.

In practice, people use whatever water supply is available to them (whether from unimproved sources or improved schemes designed with specific uses in mind, or a combination of both) for multiple purposes according to their livelihood priorities. Overlooking this reality during scheme design not only means that water may not be conveniently located or sufficient in quality or quantity for the purposes for which it is ultimately used, but can also undermine the sustainability of water supply systems. Use of systems beyond their design capacity can cause premature system failure and breakdown. There may also be a negative impact on the willingness of users to operate and maintain a system if it is not felt to cater well for their needs (Moriarty et al., 2004).

The MUS approach takes the reality of multiple uses of water as the starting point for the planning and design of new infrastructure or rehabilitation. Under MUS, water services are explicitly designed to provide water of suitable quality, quantity, and accessibility for both domestic and productive purposes. The MUS approach does not prescribe that all water needs must be met from a single source or system, but that the overall service received by users should provide for multiple uses.

Multiple-use water services can be developed by upgrading single-use systems through the implementation of add-ons, such as the addition of small drip irrigation systems to a domestic water supply system. This can be considered the 'domestic plus' entry point. Alternatively, the addition of a standpipe or washing basin to an irrigation system can be termed MUS through the 'irrigation plus' entry point. MUS can also be the result of a fully demand-based community-needs approach which matches technologies to people's multiple water requirements from the start, known as 'MUS by design'. In addition to these entry points for community-managed MUS, MUS can also emerge through self-supply.

Interest in MUS is on the rise in Ethiopia (Butterworth et al., 2011). Several implementing organizations, mainly NGOs, have developed and upgraded community-managed MUS in recent years, and MUS is now recognized as a useful service-delivery option in national policy. Households have also been building systems that serve their multiple needs for water. We will now look more closely at community-managed MUS, with self-supply discussed later in this chapter.

Community-managed multiple-use water services

One RiPPLE case study looked at domestic plus and irrigation plus MUS interventions to determine their costs and benefits (Adank et al., 2008a). The study focused on two communities: Ido Jalala and Ifa Daba, both in Goro Gutu *Woreda* in East Hararghe Zone, Oromia Region, Ethiopia. Before the project intervention, both communities used unprotected springs for multiple purposes: domestic use, watering large and small livestock, and small-scale irrigation, mainly of chat. There was, therefore, a clear demand for improved water services to meet these multiple demands.

The two communities have since taken different paths towards MUS (Figure 3.1). In Ido Jalala, a water supply system for domestic use was implemented first, and subsequently upgraded to provide water for irrigation through the construction of an irrigation canal (domestic plus pathway). In Ifa Daba the source was first developed to supply water for irrigation, but people also collected water for domestic use from the irrigation reservoir (see photo on p. 77). A standpipe for domestic water supply was connected directly to the spring later, to make it easier and safer to fetch water for domestic use: the irrigation-plus path.

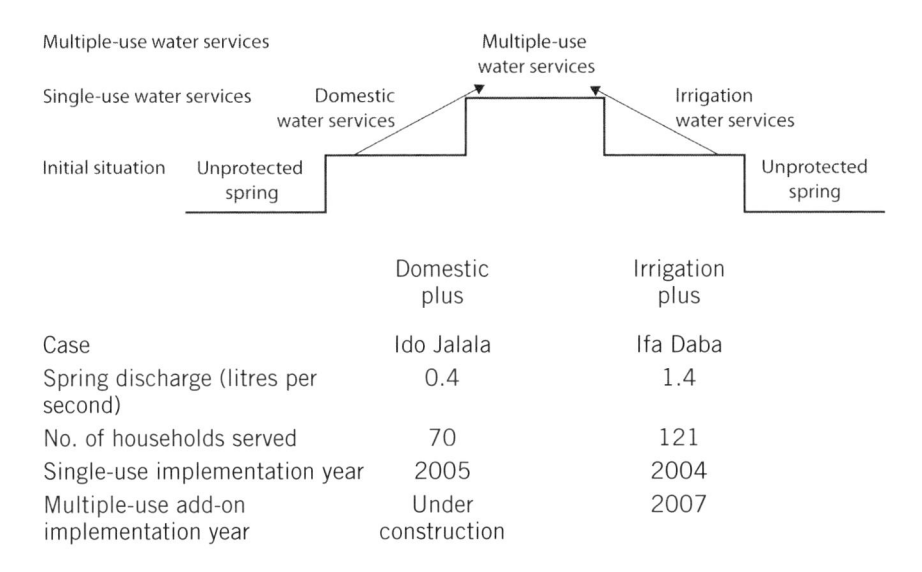

	Domestic plus	Irrigation plus
Case	Ido Jalala	Ifa Daba
Spring discharge (litres per second)	0.4	1.4
No. of households served	70	121
Single-use implementation year	2005	2004
Multiple-use add-on implementation year	Under construction	2007

Figure 3.1 Pathways towards multiple-use water services

Source: adapted from Adank et al., 2008b

In both communities, the costs of capital investment, O&M and direct support (from *woreda* and regional level) and the benefits from improved health (expenditure saved on hospital and medicine costs related to diarrhoeal deceases), time saved, and agricultural production were determined[1] (or estimated on the basis of available data). They were then compared for both the transition from unimproved sources to an improved system designed for single use, and the subsequent development of full MUS. Figure 3.2 gives an overview of the benefit/cost ratios and of the incremental benefits and costs.

76   ACHIEVING WATER SECURITY

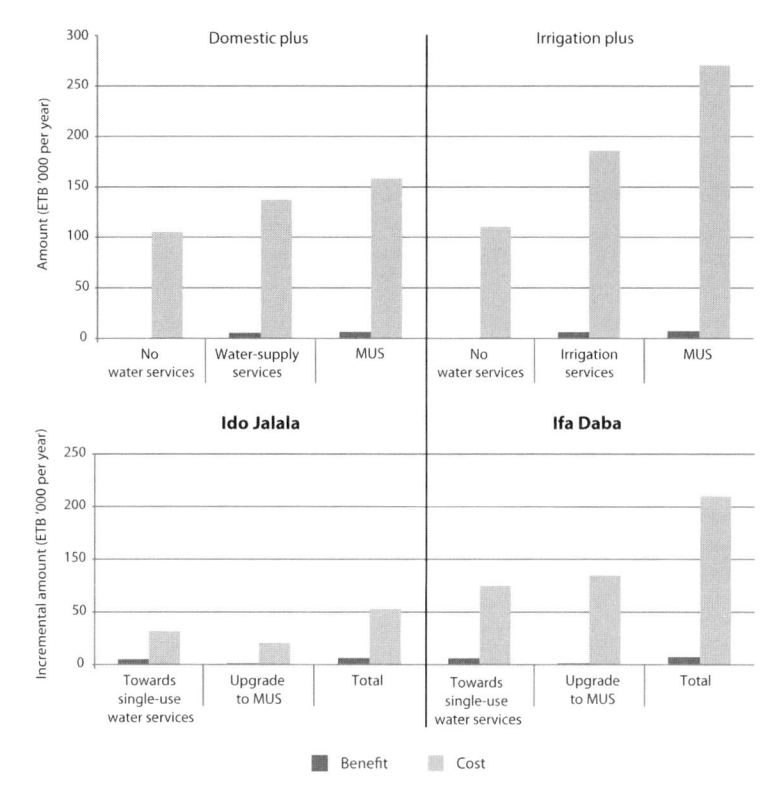

Figure 3.2   Benefit/cost and incremental benefit/cost ratios

Note: ETB standardized to year 2000 values

In Ido Jalala (domestic plus), the increase in benefits with the development of the first domestic system is attributable to measured health benefits (in line with the estimated health benefits of improved water supply determined by Hutton and Haller (2004)), and time savings. These outweigh a small decrease in benefits from irrigation caused by a decrease in irrigated area due to slightly reduced water availability for irrigation. With the addition of the irrigation canal in the step towards MUS, the irrigated area and irrigation benefits are expected to increase at the same rate as in Ifa Daba.

In Ifa Daba (irrigation plus), the initial step from the use of the unprotected spring for domestic use and irrigation, to the use of a protected spring and improved irrigation system, generated large benefits. These included an observed increase of 32 per cent in the irrigated area, and time saved as people, mostly women, spent less time collecting water. The village had only recently added a domestic standpipe, so time-saving and health benefits were estimated, rather than observed, but are thought to be in line with the measured health and time-saving benefits observed in Ido Jalala when moving from the use of an unprotected spring to the use of a domestic standpipe.

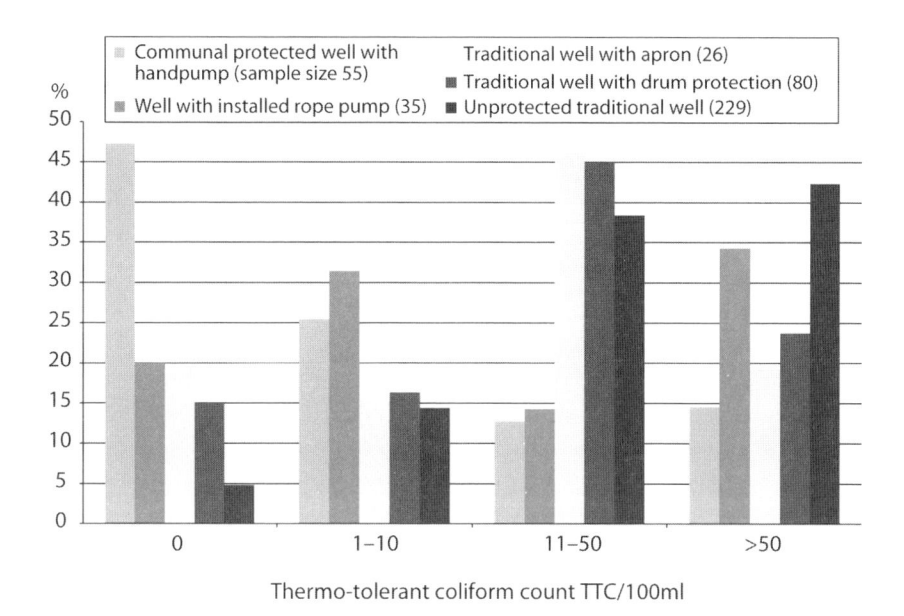

Figure 3.3 Comparison of water quality in different source types (wet season)
Source: Sutton et al., 2012

Collecting water for domestic use from the irrigation system in Ifa Daba.
Credit: Josephine Tucker

In both communities, the *incremental* benefit/cost ratio is higher for the upgrade to MUS than for the step towards single-use water services. High incremental benefits can be obtained with relatively small incremental costs when a single-use system is upgraded to cater for multiple uses. The study shows, therefore, that MUS can result in high benefits at relatively low costs.

Challenges for scaling up multiple-use water services

Although the MUS approach was highlighted in the revised UAP (MoWR, 2009), there is as yet limited evidence of scale-up.

The provision of MUS puts higher demands on organizational and institutional arrangements and intersectoral coordination and communication, for example between the Ministry of Water and Energy (MoWE) and the Ministry of Agriculture and Rural Development (MoARD) and its agencies at decentralized level. The biggest challenge to scaling up MUS approaches is likely to be slow progress in achieving more integrated planning and implementation between sectors.

Physical water availability may constrain the viability of MUS in some areas (Faal et al., 2009), but it is often the poor capacity of infrastructure rather than the lack of groundwater resources that limits MUS viability. Extending irrigation at scale may not be possible in some areas (see Chapter 6), and both the choice of technologies and local prioritization of uses must be informed by an understanding of local resource conditions and livelihoods.

Self-supply

Supporting self-supply as a service delivery approach is one way to make scarce funding resources go further (Sutton, 2009). As well as being relatively cheap because of the technologies involved, the construction and operating costs of family wells (or other technologies) are borne by households, rather than the government or its development partners.

In self-supply, the initiative and investment to build and improve individual water sources comes from individual households. This builds on widespread practice: digging family wells is common where groundwater is shallow enough, and is generally an improvement on collecting water from unprotected springs and streams, due to increased convenience and – in some cases – better water quality. Converting this unsupported *de facto* self-supply into self-supply as a service delivery model implies an increasing role for government. This involves encouraging family well construction and upgrading to ensure adequate quality of water in line with national standards for improved access. Unimproved family wells may provide very convenient water for productive uses, but need protection to be considered an improved source for drinking.

The past ten years have seen a growth in family wells and a slow growth in adoption of mechanized and rope pumps (Sutton and Hailu, 2011). However

> **Box 3.1 Recent milestones and policy developments for self-supply**
>
> - The family well campaign resulted in the construction of over 85,000 family wells in Oromia (2004–6), and almost 10,000 community hand-dug wells, highlighting demand and potential but also problems in scaling up and sustaining efforts (Mammo, 2010; UNICEF, 2010).
> - The Wolliso National Consultative Self-supply workshop in June 2008 (Anon., 2008) defined self-supply as: 'Improvement to water supplies developed largely or wholly through user investment usually at household level'.
> - The Universal Access Plan (UAP) (2009–12) drew on experiences with self-supply in other countries (MoWR, 2008), and promoted low-cost technologies at household and communal levels (MoWR, 2009).
> - The national WASH Implementation Framework (WIF) (FDRE, 2011) includes self-supply as a service-delivery model and outlines key implementation principles, including that: hardware investment should come from households with no subsidy, and government should play a supporting role, providing advice on technologies, promoting water safety, and encouraging market development (FDRE, 2011).

levels of groundwater exploitation are still well below potential in most parts of Ethiopia, and there is scope to accelerate the progress made to date in developing such sources; development which has in many cases been driven by families themselves. Where many households already use unprotected household wells which can be improved relatively easily to provide safe water for drinking, the best option for extending improved access may be to upgrade these rather than to invest in communal supply options.

The government is now promoting self-supply and creating the required environment; enabling the private sector to support self-supply and putting regulatory arrangements in place. However while self-supply offers a lower cost option for government, evidence suggests that efforts to roll out the model at scale, unless carefully designed, may face some of the same pitfalls as community management. The Oromia Family Wells Campaign (2004–6) attempted to take self-supply to scale by providing facilitation and training, but was geared to improvements that had minimal call on the household budget and was not wholly successful. Progress could not be sustained due to the need to keep investing in 'software' activities, and people did not have a strong sense of ownership of these sources, which led to poor maintenance.

Apart from leveraging household investments, there are other reasons to recognize self-supply as a complementary mode of service delivery alongside community management. It is unlikely that any single management or delivery model will be cost-effective and appropriate for all people in any given area, with varying patterns of settlement and environmental conditions. Costs also increase as hard-to-reach areas and groups are tackled. An overlapping patchwork of different models is likely to be most appropriate in the quest for universal coverage, and household family wells and rainwater harvesting have a clear role where households are scattered over large areas.

Traditional well ownership is not confined to the wealthiest. A RiPPLE survey on the use of family wells in SNNPR (largely unprotected or semi-protected) found that 58 per cent of those visited were owned by families in the lowest two quintiles in wealth ranking (61 per cent in Oromia) and one third of owners were illiterate (Sutton et al., 2011). This suggests that better supported self-supply could have a positive effect on equity of water access, although it should be noted that the protection technologies required to ensure safety for drinking would have higher costs. Neither were family wells exclusive: households were generally found to give their neighbours access to their well without charge. On average, family wells are shared by six households in SNNPR (Sutton et al., 2011) and up to 10 households in Oromia (UNICEF, 2010), with mechanized wells, which usually charge for water, shared with 20 or more households.

With family wells located close to home, their water also tends to be used for productive activities such as vegetable gardening, food processing, irrigation of seedlings. and for livestock and domestic uses. Self-supply tends to encourage multiple uses of water as decided by the owners. It may provide convenient water suitable for productive uses (e.g. market gardening or watering small livestock around the home) even where water of high enough quality for drinking cannot be obtained without more expensive technologies. Arguably, self-supply is a more scalable approach to MUS than community intervention (Butterworth et al., 2011).

Supporting productive water uses, and the development of the rural economy through private-sector initiatives, aligns well with the broad economic development model outlined in the Government of Ethiopia's Growth and Transformation Plan (GTP) (MoFED, 2010) to which the water sector and its UAP are expected to contribute.

In developing sets of water supply approaches in a given area, careful consideration should be given to their suitability for different purposes, equitable inclusion of all households, and the prospects for sustainability of community sources. The latter could be at risk if increasing numbers of households in a community 'opt out' by adopting self-supply and cease to invest in O&M of the community-level scheme. However, in practice, many households are observed to simultaneously use and value both family wells (for bulk water on their doorstep) and community sources (for perceived better water quality).

Challenges for scaling up self-supply

There are also important concerns associated with self-supply, reflected in current dialogue at national level. The key concern is the safety of drinking water from family wells. There is little data available on water quality from traditional family wells and uncertainty about water quality risks. In response,

RiPPLE surveyed family wells in SNNPR to provide new information about water quality risks and other performance measures, and to generate a more detailed understanding of how wells are developed (Sutton et al., 2011). Combined with related research supported by UNICEF (2010), this provides a better picture of traditional water supplies and their potential for low cost improvement with government support (see Box 3.2).

Not surprisingly, a comparison of different types of water sources shows that water at the source improves in quality as we move up the technology ladder from basic traditional wells to semi-protected wells, and then from family wells (with rope pumps) to protected (communal) wells fitted with handpumps. With no protection, in the wet season (worst case), 19 per cent of traditional wells had low contamination levels (<10 thermotolerant coliform (TTC) bacteria/100ml) and five per cent had zero TTC, but this rose to 34 per cent where simple measures had been taken (semi-protected wells fitted with drum or apron) to reduce the return of spilt water or run-off to the well. Fourty-four per cent of conventional community-managed handpumps achieved zero TTC (although 72 per cent were below 10 TTC/100ml). All sources need improvements, but it is important to recognize that these are the results for communal sources constructed with protection in mind, and family wells that have huge scope for low-cost improvements. These results surprised professionals who expected all traditional wells to do badly, and suggest that more could be achieved if water safety were promoted and advice provided at the lower end of the technology ladder.

Unprotected family wells show improvement in water quality during the dry season (53 per cent were found with low contamination in SNNPR, and 60 per cent of wells in Oromia were also found to provide low risk water in the dry season). This suggests that poor protection from run-off and seepage in the rainy season is a major source of contamination. Water quality was better where traditional wells were subject to a higher turnover of water through abstraction by diesel or electric pump for irrigation and home use (investigated in Oromia in the dry season), even where protection was minimal (56 per cent with zero TTC/100ml and almost 80 per cent with low contamination).

In addition, traditional wells in many areas provide water for more days per year than conventional handpumps, compensating in part for their lower water quality. Some 90 per cent of wells in four *woredas* had not dried up in the previous five years. Those that were less reliable tended to be in areas with poorly consolidated ground and no tradition of lining the shaft. Here, the introduction of low-cost lining techniques would improve the situation.

Few traditional wells had proper well-protected headworks to prevent the return of dirty water to the well. They were, at best, semi-protected. Traditional well owners have had little advice on simple measures of protection, lining and water lifting, and almost all wanted technical advice, suggesting considerable scope for improvement. Water quality results reflect the effects of poor site hygiene and, in some cases, poor installation design or practice for all wells. Improved training for artisans and the promotion of hygiene education amongst well owners and users could deliver considerable improvements in the quality of water provided through community and household supplies. Figure 3.3 gives more detail on water quality from sources of different types and with varying levels of protection.

Perhaps as a result of these concerns, and despite stated policy intentions, implementation of the self-supply approach has, like MUS, lacked a clear model or strategy until very recently. It has not been possible to develop models to accelerate and improve family well construction and use despite the UAP 2009 policy. One problem is that budgets (such as UAP plans) focus on capital investments in new construction, leaving little incentive for *woredas* and regions to implement self-supply as this would reduce their claim for funds from zonal and regional finance offices. There is no mechanism yet for regions and *woredas* to request funding for self-supply supporting activities, even though they might be more cost-effective in generating coverage than new capital investments in community-managed water supply.

It is therefore difficult to make the case for such software interventions, which is a major concern. The government's role in self-supply is to establish the right enabling environment for households to invest: creating the conditions to accelerate family construction of wells (e.g. enabling the development of a private sector to support such construction), ensuring protection standards where wells are to be used for drinking, and promoting practices for their safe use. Opportunities to link self-supply to prevailing community management interventions in training, monitoring, and promotion are little developed, as are potential links with agriculture and health (self-supply requires a household-based approach to sanitation and hygiene interventions, and good environmental sanitation to protect shallow groundwater from contamination, especially if use of agrochemicals increases in the future).

Much could be done by integrating training for self-supply support services and promotion into existing training opportunities for masons, staff from water, health and agriculture extension offices and WASHCOs. The MoWE is developing a Self-Supply Acceleration Programme to address these issues: a response, in part, to research on this issue by RiPPLE and its partners.

Another disincentive has been that the contribution of self-supply has not been captured in sector monitoring. Promotion of self-supply at scale has stalled partly because such sources were not counted during coverage monitoring. This explains, in part, why huge strides in developing access through family wells in Oromia, for example (Mammo, 2010), were not built upon or sustained. Since coverage is based on the numbers of improved *community-managed* systems, new family wells were not, according to the statistics, improving access.

The new National WASH Inventory (NWI) (see Chapter 2), included a question in its household survey on the primary household drinking water source, which could include family wells (MoWE, 2011b). This will yield important new information on genuine access to water, but is unlikely to reflect the true density of family wells, as few will be included in the inventory. Only wells used as the main source of drinking water will be captured, yet many others may exist with potential for improvement, and may currently be used for drinking in conjunction with other sources. There is, to date, no agreement on which family wells should be considered as safe sources, and therefore included in coverage. The inclusion of some family wells in the NWI paves the way for greater recognition of their importance in the future. However an acceptable benchmark needs to be established to ensure that only family wells which provide water of appropriate quality for drinking are counted as increasing coverage of improved services.

Ethiopia's policy environment is highly dynamic as the country refines its approaches, and current policies and plans are not altogether consistent on self-supply. The updated Universal Access Plan (UAP 2)[2] does not yet highlight self-supply, even though this is one way to strengthen the link between WASH and economic development as set out in the GTP. However the UAP 2 is now being reviewed and the revised version is expected to include self-supply. The new WASH implementation framework (WIF) (FDRE, 2011) does identify self-supply as a service delivery model alongside *woreda*-managed projects (those constructed by *woreda* government and handed over to communities for management as per the traditional community-managed model) and community management projects (projects with community-managed grants for development of sources). The framework also sets out some key principles for how this should be done (see Chapter 1).

The lack of information on the de facto self-supply that already exists, and limited piloting of approaches for promoting, supporting, and regulating self-supply (beyond simply piloting technology options), means that there are no guidelines on how to establish a more enabling environment for self-supply. There is limited systematic encouragement or sustained support for self-supply, with some localized exceptions such as the Productive Safety Net Programme (PSNP) in parts of SNNPR. In Oromia, rates of construction tailed off after the campaign for family well digging, indicating the importance of basing support programmes on real household demand rather than attempting to impose a one-size-fits-all solution.

Multiple-use water services through self-supply

Family wells are often used for multiple uses such as drinking water, water for livestock, and crop irrigation, as confirmed by an inventory of water sources and water use carried out by RiPPLE in Mirab Abaya *Woreda* in SNNPR (Abebe et al., 2010). During the survey, well owners identified many benefits related to multiple uses of water from the family well, including increased food security, health, school attendance, and better childcare. More easily accessible well water generated major economic changes with increased animal watering (around 90 per cent of family wells in SNNPR are used for livestock) and crop production (traditional wells are used for irrigation in 20–30 per cent of cases, with rope pumps and mechanized wells used for irrigation in 43 and 68 per cent of cases respectively) (Sutton et al., 2011).

Households implementing their own systems are not constrained by an external 'sectoral focus', but implement structures that work for them and meet their multiple demands, demonstrating a clear link between self-supply and MUS.

As noted, all wealth groups make multiple use of water from family wells. However, the wealthy, with access to assets other than water (notably land, capital, livestock, and household labour), are better able to capitalize on multiple-use potential (Box 3.2)

Box 3.2 Self-supply and multiple uses in Mirab Abaya

In the highland community of Weye Barena in Mirab Abaya *Woreda*, people from all wealth groups were growing vegetables using water from family wells. Some better-off households also used hand-dug wells and un-developed spring water sources to water apple seedlings. Vegetables are used primarily for household consumption, while apple seedlings are sold for Ethiopian birr (ETB) 40 each (US$3.11).[3] Some farmers were able to use this additional income to buy extra cattle and improve their livelihoods. The poorest farmers, however, do not plant apple trees as this does not satisfy their immediate food needs.

Source: Abebe et al., 2010

Conclusion

The functionality and sustainability of community-managed water services need to be improved to achieve universal access to water services. The following recommendations focus on the possibilities to improve service levels and sustainability by adopting MUS or self-supply approaches to complement existing approaches to water supply. More general strategies for supporting community-managed services have been widely discussed elsewhere (see Chapter 5).

As well as obvious improvements needed in water provision and support to water providers, ensuring that community-managed water supply responds to people's multiple demands can increase community willingness to maintain systems, ensuring sustainable water supply. Applying an MUS approach can improve the economic status of the users, which, when well-managed, can also improve the financial sustainability of community-managed water supply.

It is recommended, therefore, that planners and implementers give far more consideration to people's multiple water demands in the planning and implementation of new systems, or the upgrading of existing single-use systems. It is vital to create awareness and build capacity among planners and implementers in the area of MUS.

Obviously the provision of MUS may sometimes be limited by a lack of natural water resources, but it should not be limited by the sub-sectoral bias of planners and implementers. There is a need for better integration and coordination between different sub-sectors, including watershed protection, water supply, health, irrigation, and livestock, in both policy and practice. This will require additional resources and capacities at all levels.

Promoting self-supply to improve service and increase coverage also requires different roles and strategies at all levels of public service. Household investment in water supply is relatively slow at present and the enabling environment is not as conducive as it could be. The best way to accelerate self-supply is to pilot the self-supply approach to establish what works well and what does not for different areas and household groups, with a focus on how the supporting software (awareness raising, planning, technical support, monitoring, access to credit) can be delivered by government to support household investment in hardware and private-sector provision of services, and ensure a minimum standard of water quality.

Community-based management and self-supply are complementary, especially when applying an MUS approach. People in rural areas often depend on multiple water sources to satisfy their different needs, obtaining small quantities from community-managed sources for drinking and cooking, while using easily accessible water from family wells and other unimproved sources for other domestic and productive uses. Water services must be addressed holistically, taking people's needs and capacities as a starting point, and allowing for use of different sources for different purposes to fulfil their varied water needs.

Notes

1. Figures based on in-depth interviews and focus group discussions with households of different wealth levels; interviews with project staff members, water committee members, and *woreda* support staff; examination of water committee accounts, project documents, and *woreda* budgets; data on disease incidence from health clinics; data on agricultural prices and inputs from the *woreda* Office of Agriculture; observation and measurement of source discharge, water quality, reliability, accessibility, and use; and monitoring by scheme caretakers of investments (money, time, other resources) in O&M.
2. The Universal Access Plan for rural water, 2005 (UAP 1) was revised in 2009 and again in April 2011, now called UAP 2. It is realized by the MoWE that even the UAP 2 is not adequately addressing the principles of the WIF. The UAP 2 (rural water) was made to focus on securing water supply access to a total of 18 million rural people in 2011–15 designed with a target to reach 98 per cent rural water supply access by 2015, at the implementation rate of 7 per cent per annum and a cost of US$1.7 billion.
3. Exchange rate ETB 1 = US$0.08 (1 January 2010)

References

Abebe, H. and Deneke, I. (2008) 'The sustainability of water supply schemes: a case study in Alaba Special Woreda', *RiPPLE Working Paper 5*, RiPPLE, Addis Ababa. All RiPPLE papers available from: <www.rippleethiopia.org> [accessed July 2012].

Abebe, H., Bedru, M., Ashine, A., Hilemariam, G., Haile, B., Demtse, D., and Adank, M. (2010) 'Equitable water service for multiple uses: a case from Southern Nations, Nationalities, and People's Region (SNNPR), Ethiopia', *RiPPLE Working Paper 17*, RiPPLE, Addis Ababa.

Adank, M., Jeths, M., Belete, B., Chaka, S., Lema, Z., Tamiru, D., and Abebe, Z. (2008a) 'The costs and benefits of multiple uses of water: the case of Gorogutu Woreda of East Hararghe Zone, Oromiya Regional States, eastern Ethiopia', *RiPPLE Working Paper 4*, RiPPLE, Addis Ababa.

Adank, M., Belete, B., and Jeths, M. (2008b) 'Costs and benefits of multiple uses of water: a case from Ethiopia', International Symposium of Multiple-use Water Services, 4–6 November 2008, Addis Ababa.

Anon (2008) *National Consultative Workshop on Self Supply: Report on Presentations, Discussions and Follow up Actions, 4–6 June 2008 Workshop*, Wolisso, Ethiopia. Ministry of Water Resources (MoWR) / United Nation's Children's Fund (UNICEF) / Water and Sanitation Programme [website]. Available from: <www.rwsn.ch/> [accessed July 2012].

Butterworth, J., Visscher, J.T., Steenbergen, F.V., and Koppen, B.V. (2011) 'Multiple use water services in Ethiopia scoping study', *International Water Management Institute (IWMI)/IRC International Water and Sanitation Centre Report for Rockefeller Foundation*, IWMI/IRC, The Hague.

Chaka, T., Yirgu, L., Abebe, Z., and Butterworth, J. (2011) *Ethiopia: Lessons for Rural Water Supply; Assessing Progress towards Sustainable Service Delivery*, IRC, The Hague.

Deneke, I. and Abebe, H. (2008) 'The sustainability of water supply schemes: a case study in Mirab Abaya woreda, Hawassa', *RiPPLE Working Paper 4*, RiPPLE, Addis Ababa.

Faal, J. Nicol, A., Tucker, J. (2009) 'Multiple-use water services (MUS): cost-effective water investments to reduce poverty and address all the MDGs', *RiPPLE Briefing Note*, RiPPLE, Addis Ababa.

Federal Democratic Republic of Ethiopia (FDRE) (2011) *The WASH Implementation Framework (WIF) – Summary*, version: 27 July 2011, FDRE, Addis Ababa.

Hutton, G. and Haller, L. (2004) 'Evaluation of the costs and benefits of water and sanitation improvements at the global level', World Health Organization (WHO), Geneva. Available from: <www.who.int/water_sanitation_health/wsh0404.pdf> [accessed July 2012].

Lockwood, H. and Smits, S. (2011) *Supporting Rural Water Supply: Moving Towards a Service Delivery Approach*, IRC, The Hague and Aguaconsult. Available from: <www.waterservicesthatlast.org/Resources/Multi-country-synthesis> [accessed July 2012].

Mammo, A. (2010) 'Assessment of local manufacturing capacity for rope pumps in Ethiopia', unpublished draft report, December 2010, prepared for UNICEF.

Ministry of Finance and Economic Development (MoFED) (2010) *Growth and Transformation Plan (GTP) Draft Report*, Federal Democratic Republic of Ethiopia (FDRE), Addis Ababa. Available from: <www.ethiopians.com/Ethiopia_GTP_2015.pdf> [accessed July 2012].

Moriarty, P., Jeths, M., Abebe, H., and Deneke, I. (2009) 'Reaching universal access: Ethiopia's Universal Access Plan in the Southern Nations, Nationalities, and People's Region (SNNPR), A synthesis paper of recent research under the RiPPLE Programme's Governance and Planning theme', *RiPPLE Synthesis Paper*, RiPPLE, Addis Ababa.

Moriarty, P., Butterworth, J., van Koppen, B. (2004) 'Beyond domestic: case studies on poverty and productive uses of water at the household level', *Technical Paper Series, No. 14*, IRC, Delft. Available from: <www.irc.nl/page/6129> [accessed July 2012].

Ministry of Water and Energy (MoWE) (2011a) *Revised UAP (Rural Water Supply) Final Report*, FDRE, Addis Ababa.

MoWE (2011b) *National Water Supply, Sanitation and Hygiene Inventory*. Unpublished formats for data collection, FDRE, Addis Ababa.

Ministry of Water Resources (MoWR) (2008) *Reformulation of Strategies and Plans for an Accelerated Implementation of the Universal Access Plan for Rural Water Supply*, FDRE, Addis Ababa.

MoWR (2009) *Review of Rural Water Supply Universal Access Plan, Implementation and Reformulation of Plans and Strategies for Accelerated Implementation*, FDRE, Addis Ababa.

Rural Water Supply Network (RWSN) (2012) *Handpump Data Collated by Peter Harvey, UNICEF Zambia, May 2007* [website]. Available from: <www.rwsn.ch/prarticle.2005-10-25.9856177177/prarticle.2005-10-26.9228452953/prarticle.2009-03-09.1365462467> [accessed July 2012].

Sutton, S. (2004) 'Self supply: a fresh approach to water for rural populations', *WSP Field Note*, Water and Sanitation Programme (WSP-)Africa and World Bank, Nairobi.

Sutton, S. (2007) 'An introduction to self-supply. Putting the user first: incremental improvements and private investment in rural water supply', *WSP Field Note*, Rural Water Supply Network, Self-Supply Flagship, Nairobi.

Sutton, S. (2009) 'An introduction to self-supply. Putting the user first: incremental improvements and private investment in rural water supply', *WSP Field Note*, Rural Water Supply Series, WSP–Africa and World Bank, Nairobi.

Sutton, S. and Hailu, T. (2011) 'Introduction of the rope pump in SNNPR, and its wider implications', *RiPPLE Working Paper 22*, RiPPLE, Addis Ababa.

Sutton, S., Mamo, E., Butterworth, J., and Dimtse, D. (2011) 'Towards the Ethiopian goal of universal access to rural water: understanding the potential contribution of self supply', *RiPPLE Working Paper 23*, RiPPLE, Addis Ababa.

Sutton, S., Butterworth, J., and Mekonta, L. (2012) *A Hidden Resource: Household-led Rural Water Supply in Ethiopia*, IRC, The Hague.

Tekalign, T. (2001) *Evaluation of Financial Sustainability of Hitosa and Gonde-Iteya Water Supply Schemes*. WaterAid, Addis Ababa.

UNICEF (2010) *Benchmarking for Self Supply (Family Wells)*, unpublished report (December 2010). UNICEF, Addis Ababa.

Van Koppen, B., Smits, S., Moriarty, P., Penning de Vries, F., Mikhail, M., and Boelee, E. (2009) *Climbing the Water Ladder – Multiple-Use Water Services for Poverty Reduction*, IRC, The Hague.

Whittington, D., Davis, J., Prokopy, L., Komives, K., Thorsten, R., Lukacs, H., Bakalian, A., and Wakeman, W. (2008) 'How well is the demand-driven, community management model for rural water supply systems doing? Evidence from Bolivia, Peru, and Ghana', *Brooks World Poverty Institute (BWPI) Working Paper 22*, BWPI, Manchester. Available from: <www.bwpi.manchester.ac.uk/resources/Working-Papers/bwpi-wp-2208.pdf> [accessed July 2012].

About the authors

Marieke Adank has been working with the IRC International Water and Sanitation Centre for almost a decade. Her work is mainly focused on Africa, specifically Ethiopia and Ghana. With her background in irrigation and water engineering, her interest and expertise in multiple-use water services will not come as a surprise. Besides this, she focuses on water governance in rural and urban areas, with a special interest in small towns.

John Butterworth is a Senior Programme Officer at the IRC International Water and Sanitation Centre where he coordinates IRC's Ethiopia Country Programme. He managed IRC's inputs to the RiPPLE project, and coordinated research activities on behalf of the consortium on sector monitoring and self-supply.

Sally Sutton's first encounters with the technical and social complexities of rural water supply were during four years' research in rural Oman in the early 1970s. Ten years of practical hydrogeology for international consulting engineers were then followed by working for the Zambian Government on a range of approaches to rural water supply including the beginnings of community- and household-led initiatives (self-supply). Since that time she has worked as an independent consultant throughout Africa, on planning and evaluation of water and sanitation programmes and developing the acceleration of self-supply in several countries as coordinator of the Self Supply Theme of the Rural Water Supply Network (RWSN) from 2003–11.

Zemede Abebe is a Programme Advisor for RiPPLE and the Hararghe Catholic Secretariat in Ethiopia, and is also leading the Disaster Risk Reduction (DRR) Team for the Catholic Relief Service (CRS) in Eastern Africa/South Sudan. With expertise ranging from water resources management to agro-enterprise and building learning alliances, Zemede has over 15 years' experience working in a range of research, emergency, rehabilitation, and development programmes. He holds an MSc in Agricultural Economics from Haramaya University, Ethiopia.

CHAPTER 4
Sanitation and hygiene promotion in rural communities: the Health Extension Programme

Peter Newborne and Anu Liisanantti

The initial enthusiasm for government health extension programmes after the 1978 Declaration of Alma-Ata has waned in much of sub-Saharan Africa (SSA), since the 1980s. The Health Extension Programme (HEP) in Ethiopia, which started in 2002, is one exception. This national initiative to provide preventive as well as curative primary health care is driven by the Ethiopian Government and includes the promotion of sanitation and hygiene (S&H) through the deployment of salaried health extension workers and voluntary community health promoters. The regional government of the Southern Nations, Nationalities and People's Region (SNNPR) has led a strong campaign to promote latrine construction and improved hygiene in line with the HEP. Research led by the Research-inspired Policy and Practice Learning in Ethiopia and the Nile Region (RiPPLE) programme in 2008 and 2010 concluded that the HEP frontline health workers played a key role in motivating rural households to construct latrines and improve hygiene. While lead responsibility for the HEP remains with the Ethiopian Government, more external funding is needed to meet national targets for sanitation and infant/child mortality. Donors and NGOs should help to fill the HEP resource gaps, avoiding parallel health extension programmes that by-pass a government initiative that is already on its way to becoming a model for other countries in the region.

Introduction

Sanitation specialists have reiterated (Mara et al., 2010) the importance of including S&H within appropriately institutionalized and adequately resourced health extension systems. The 1978 Declaration of Alma-Ata aimed to make primary health an integral part of national health systems, with one of its eight key elements being 'an adequate supply of safe water and basic sanitation' (Article VII.3). This reflects the key role of improved S&H practice in preventing tropical diseases and mortality from diarrhoea (Mara et al., 2010). In recent decades, however, primary health care has been side-lined by health planners and managers in many countries (Lawn et al., 2008). Most senior decision-makers in health administrations have medical training, and health services are frequently 'medicalized' (Rehfuess et al., 2009; Newborne and Samuels, 2010). Health is seen to be predominantly curative, about 'giving out drugs' (ibid.). Preventive aspects are relatively neglected, with governments playing down, for example, the responsibility of health administrations to provide leadership on S&H.

In terms of their content, government, donor, and NGO programmes in S&H in SSA, including Ethiopia, are converging. Core principles have received broad acceptance, such as the move away from systematic subsidies for latrine construction with gifts of hardware (e.g. latrine slabs) towards support to hygiene promotion, through the 'software' roles of promoters and facilitators.

The Health Extension Programme (HEP), first introduced in Ethiopia in 2002, is a notable exception to the otherwise observed trend of neglect of preventive measures. The federal Ministry of Health (MoH) has made prevention an explicit part of its portfolio of primary health care, including S&H, and continues to lead the process of building the HEP as a national initiative.

To place the HEP in a broader context, we compare examples of health extension programmes in other SSA countries. The focus here is on government-led programmes, as compared with individual initiatives of donors or NGOs in selected locations (UNICEF, 2004). The responsibilities of governments extend across national territories, and communities look to public health authorities to provide the *leadership* for the resources (public or private) to be mobilized to reach them. A critical issue, discussed below, is the extent of donor and NGO support to the implementation of the HEP, including the promotion of S&H.

This chapter brings together the findings of two studies led by Research-inspired Policy and Practice Learning in Ethiopia and the Nile Region (RiPPLE) on how S&H has been promoted under the HEP in the Southern Nations, Nationalities, and People's Region (SNNPR) of Ethiopia where the regional government has displayed strong leadership.

Sanitation and hygiene in Ethiopia: status and targets

There are diverging views on the extent of the S&H challenge in Ethiopia. According to government figures, about 60 per cent of the total population now has access to sanitation facilities (56 per cent in rural areas), although only 20 per cent of households actually utilize latrines (MoH, 2010). Estimates published internationally are, however, much lower. According to information supplied by the Ethiopian Government, the 2012 report of the World Health Organization (WHO)/United Nations Children's Fund (UNICEF) Joint Monitoring Programme (JMP) estimated rural sanitation access, including basic and shared facilities, at 47 per cent in 2010, up from 29 per cent in 2008 (WHO/UNICEF, 2010; 2012). The Ethiopian WASH Inventory that is being conducted at present (see Chapter 2) is intended to resolve the difference and the results are awaited. The inventory is expected to confirm the substantial variations in access to S&H between different regions of Ethiopia. Access to latrines, for example, varies significantly between regions with some regions reporting up to 75 per cent access, while mainly pastoralist regions (Afar, Somali, Gambella) report access rates of 10 per cent or less (MoH, 2011).

In 2005, child mortality in Ethiopia was recorded as being among the highest in the world, with nearly one in every ten babies born (97 per 1,000) not reaching their first birthday, and one in every six children dying before the age of five (World Bank, 2005). The situation has improved according to the MoH, with the infant mortality rate (IMR) now reduced to 77 deaths per 1,000 live births and the under-five mortality rate (U5MR) to 123 per 1,000. The MoH has set a target to reduce IMR to 45 deaths per 1,000 live births and the U5MR to 85 per 1,000 by 2015.

In the 2011–15 Strategic Action Plan, the MoH has confirmed its goal of 100 per cent coverage in basic sanitation, and 84 per cent access to improved sanitation by 2015. The definition of basic sanitation employed in the Strategic Action Plan is the 'lowest-cost option for securing sustainable access to safe, hygienic and convenient facilities and services for excreta and sullage disposal that provide privacy and dignity while at the same time ensuring a clean and healthful living environment, both at home and in the neighbourhood of users' (MoH, 2011: 5). Improved sanitation is defined as 'the process where people demand, develop and sustain a hygienic and healthy environment for themselves by erecting barriers to prevent the transmission of diseases, primarily from faecal contamination' (MoH, 2011: 5).

As part of the Health Sector Development Programme III (MoH Health Sector Strategic Plan, HSDP-III), the National Hygiene and Sanitation Strategy and the National Protocol for Hygiene have been developed, and implementation of Community-led Total Sanitation (CLTS) has begun. CLTS approaches were introduced in 2006 and are now being expanded to include an 'H' for hygiene promotion. CLTSH focuses on facilitating a collective approach to communal sanitation behaviour change rather than simply constructing toilets. National CLTSH guidelines are currently being produced, focusing on good quality facilitation, follow-up. and ensuring the necessary supply streams are in place. According to the MoH, in 2010 CLTSH has been facilitated in 102 *woredas* (districts) and 4,643 villages, of which 1,913 have declared themselves open defecation free (MoH, 2011).

Before 2003, SNNPR had one of the lowest S&H coverage levels in the country, recorded (according to official figures) at just 16 per cent (BoH, 2006). The extent of the regional budget allocated to S&H was also amongst the lowest, at only 0.4 per cent of the health budget (Shiferaw and Mariam, 2003). The scope of education on S&H was limited, with a lack of appropriate strategies for community education and mobilization (ibid, 2003). Messages on S&H were communicated when community members came to health institutions to obtain health services. The approach to S&H was supply-driven, with health authorities raising the expectations of households that the government would give them incentives such as hand-outs (discussed below) to improve S&H practices. As a result, the regional government recorded low household demand for S&H services.

The Health Extension Programme in Ethiopia

The HEP is a prominent example of a government-led, national initiative for the provision of primary health-care services. It combines a wide range of services to form an integrated community health package, including seven hygiene and environmental sanitation components (Table 4.1).

Table 4.1 Sanitation and hygiene – elements of the Health Extension Programme

1 Hygiene and Environmental Sanitation	3 Family Health Service
• *Excreta disposal*[1] • Solid and liquid waste disposal • *Water quality control*[1] • Food hygiene • Proper housing • Arthropods and rodent control • *Personal hygiene*[1]	• Maternal and child health • Family planning • Immunization • Adolescent reproductive health • Nutrition
2 Disease Prevention and Control	4 Health Education and Communication
• HIV/AIDS and other sexually transmitted diseases prevention and control • Tuberculosis (TB) prevention and control • Malaria prevention and control • First aid	• Health education • Communication

1 The three S&H elements referred to specifically in this chapter
Source: adapted from MoH, 2007

Given that hygiene is dependent on water access, e.g. for handwashing, S&H interventions need to be coordinated with water investments, which requires collaboration between government health and water agencies. This chapter focuses on three S&H elements – excreta disposal, water quality control, and personal hygiene – and how they are delivered as part of the HEP by the health extension workers (HEWs) and community health promoters (CHPs) deployed in rural communities.

The HEWs are salaried employees of the *woreda* (district) health authorities, while the CHPs are unpaid volunteers from communities. In developing countries, where budget constraints mean that government health services have too few employed personnel, the engagement of community members to provide basic health services has been identified as an extension strategy.

Each of the three S&H elements highlighted in Table 4.1 – human excreta disposal via latrines, handwashing, and safe water storage and handling in the home – aim to prevent ill-health. Curative measures, such as provision of medicines, are carried out under other components of the HEP.

The central philosophy of the HEP is that, if the right knowledge and skills are transferred, households can take responsibility for improving and maintaining their own health (MoH, 2011). The approach of the 2002 HEP was confirmed in the National Hygiene and Sanitation Strategy of 2005, and is renewed in the 2011–15 Strategic Action Plan. By September 2010, more than 34,000 HEWs had been recruited across the country to impart the necessary knowledge and skills (MoH, 2011).

The SNNPR experience in sanitation and health promotion post-2003

The Bureau of Health (BoH) of the SNNPR Regional Government began its new community health initiative in 2003, in line with the HEP. The approach developed by the BoH attracted international attention in a field note of the Water and Sanitation Programme (WSP) (Bibby and Knapp, 2007). With the agreement of the BoH, the RiPPLE programme selected this approach for detailed research in 2007/8, with a further study in 2009/10 (Tefera, 2008; Terefe and Welle, 2008; Newborne and Smet, 2008; and Behailu et al. 2010).

One important element of the SNNPR BoH's approach was the promotion of basic latrine construction and improvement of hygiene practices – again, in line with the HEP. In accordance with the HEP's demand-led philosophy, the new strategy emphasized raising the awareness of households on S&H and encouraging each household to take responsibility for action.

Hardware subsidies were no longer provided. Households were to start from simple, traditional pit latrines, constructed from locally available materials and, subsequently, upgrade their standard as awareness grew and opportunity allowed. The aim was that households would move up the 'sanitation ladder' as illustrated in Figure 4.1.

The approach aimed to reach community members via the HEWs and CHPs who were deployed progressively within the region. Targeted at households, the BoH strategy was not 'community-led' in the manner of CLTS as while the BoH approach had some similar features (e.g. no hardware subsidy), CLTS was a later introduction to Ethiopia, from 2006 onwards, as discussed below.

The first step in the RiPPLE-led research in 2007/8 was to study the policy-making process that launched the S&H strategy in SNNPR. A combination of political promotion and institutional mobilization was successful in launching and rolling-out the BoH strategy (Terefe and Welle, 2008). The placing of S&H as part of the basic community health package was designed to be politically attractive – a so-called 'broad-based, low cost and high impact' oriented approach – and financially and administratively feasible (ibid: 3). 'Ignition' documents were written with a strong communication orientation

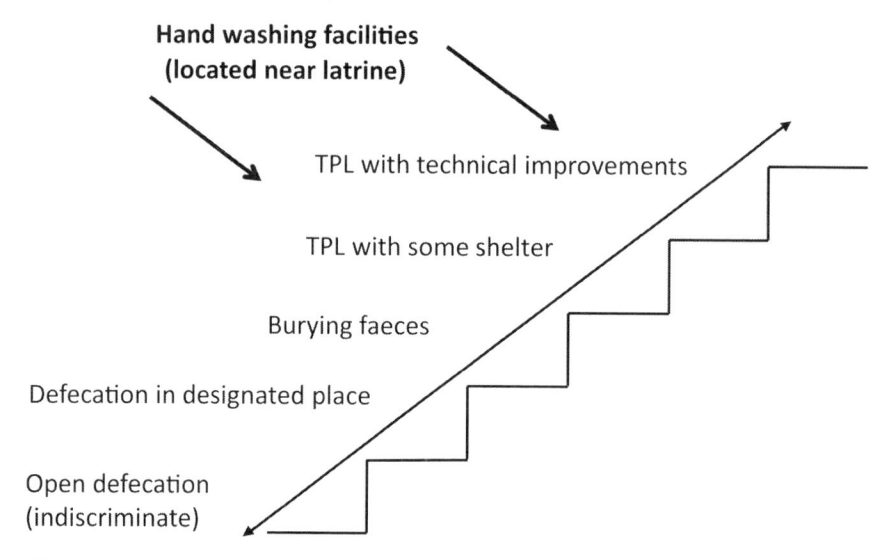

```
                    Hand washing facilities
                     (located near latrine)

                              TPL with technical improvements

                              TPL with some shelter

                              Burying faeces

         Defecation in designated place

     Open defecation
     (indiscriminate)
```

Figure 4.1   A typical 'sanitation ladder' in rural SNNPR

Note: TPL = traditional pit latrine

Source: Haddis, 2008

to persuade politicians, motivate civil servants, and build consensus for action by a range of stakeholders (ibid.). Implementation tools, piloted by donors in the region, were selected opportunistically by the BoH, and some donor funds were leveraged for software aspects (ibid.).

The RiPPLE research team also surveyed latrine construction and use, handwashing, and water storage/handling by rural households in six *kebeles* (communities) in two districts of SNNPR, Halaba Special *Woreda* and Mirab Abaya *Woreda*. The RiPPLE study aimed to identify what progress had been made and how, as an input into the SNNPR Learning and Practice Alliance (LPA), established in May 2007 with RiPPLE support.

The research highlighted the following aspects of the S&H strategy in SNNPR and issues arising from its implementation:

- **Latrine construction**: the survey found a substantial increase in just a few years in the number of household latrines using basic technology in both districts (Newborne and Smet, 2008). Some questions arose on the sustainability of this wave of latrine construction in prevailing environmental conditions (Tefera, 2008).
- **Hygiene promotion**: on hygiene, the survey and interviews suggested there was some increase in awareness, although observation pointed to continued poor practices in handwashing and water storage/handling, suggesting a need for a stronger focus on household behaviour change (Newborne and Smet, 2008).

- **Household motivation, first phase**: in the first years of roll-out of the S&H strategy, 2003–5, the 'command' aspect of the *kebele* authorities in the two districts was the key driving force behind latrine construction. *Kebeles* have authority as part of the formal institutional (and political) hierarchy – the 'obeying of orders from above' being a common feature of social interaction in Ethiopia (Vaughan and Tronvoll, 2003: 33). *Kebele* chairs and other *kebele* cabinet members communicated the community health policy downwards with other government policies. Households built latrines, but, in the absence of adequate technical support to guide construction, the latrines were not often used.
- **Training**: at this stage, the CHPs were receiving only two days of training, and technical support was required from HEWs to convert that training into effective CHP practice.
- **Household motivation, second phase**: in the second phase, post-2005, promotion, rather than command, was applied to encourage households to make changes to S&H facilities and practices. *Kebele* influence still played its part, but *kebele* staff shifted their role to enable the work of the CHPs and HEWs, with *kebele* officials still aware of the political importance accorded to the S&H as part of the HEP. HEWs also indicated that they sometimes relied on the authority of the *kebele* to ensure that households were constructing latrines. This second phase saw increased rates of latrine construction by households.
- **Messaging**: from 2005 onwards, the deployment of the trained HEWs in each *kebele* raised the level of communication on S&H, including technical input on construction of latrines. The engagement of CHPs raised the number of community-level communicators significantly from one CHP for every 2,000 households in a *kebele* to 12–28 CHPs in a *kebele*, or one CHP per around every 50 households. Household representatives responding to the RiPPLE survey stated that their primary sources of S&H messages were CHPs and HEWs: 35 per cent of households indicated that they heard about S&H messages for the first time from CHPs, 31 per cent cited HEWs, and 25 per cent cited *kebele* officials.

The 2008 study concluded that the HEWs and CHPs, as the frontline health workers, had a growing significance in the success of the S&H strategy, acting as promotional change agents to motivate the household construction of latrines.

As for application of the BoH policy by enforcement – the 'stick' alongside the 'carrot' (Mara et al., 2010) – only a small minority of respondents in both *woredas* said that they had built their latrine out of fear of punishment: the application of sanctions on non-compliance seems to have been exerted sparingly. The regional Public Health Proclamation of 2004 had decreed that failure to construct sanitation facilities was punishable by law, but the lack of by-laws under the Proclamation meant that there was no uniform system of sanctions, e.g. fines, for open defecation and urination. Officials in *kebeles* applied their authority in the manner each thought appropriate, such as threats of punishment, but actual enforcement by *kebeles* appears to have been limited.

The role of health extension workers and community health promoters in sanitation and hygiene promotion

The second RiPPLE study in 2009/2010 focused more closely on the roles of HEWs and CHPs in relation to promotion of S&H in rural communities, as seen through their activities in the same two *woredas* (Halaba Special and Mirab Abaya) in SNNPR.

The study confirmed that the work of HEWs and CHPs in the two sample *woredas* was clearly valued by communities, but that there was a need to strengthen certain aspects of the operation of S&H elements within the HEP (Behailu et al., 2010):

- **Training and supervision**: it was recommended that the BoH review existing measures to build the capacity (knowledge and skills) of HEWs and CHPs, and plan for more training of both. HEWs were lacking supervision and support from *woreda* health offices, and, in turn, HEWs needed to organize more group meetings with CHPs (twice-monthly) to guide them in the planning and organization of their work, and collaborate in reviewing the means of service delivery under the HEP.
- **Outreach**: the HEWs considered that the geographical coverage of supervision was inconsistent: remote *kebeles* did not generally receive as much attention as those close to *woreda* health offices and health centres; the natural consequence of this, they thought, would be weakening of the outreach of the HEP to remote areas in SNNPR (a poor region, whose more remote rural communities already suffer from marginalization).
- **Motivation of front-line health workers/promoters**: it was felt that retaining the services of HEWs, and especially CHPs, was becoming more difficult: a system of rewards for good performance had not yet been established and some kind of incentive system was needed.
- **Equipment and materials**: it was thought that one way to motivate HEWs and CHPs would be provision of basic equipment, like umbrellas and materials for information, education, and communication (IEC), as well as means of transport (for example, a bicycle for each HEW and CHP). There was a thought to be a need to construct health posts in the *kebeles* that lacked this facility.
- **Messaging**: it was recommended that arguments on dignity and privacy should be communicated to persuade households to adopt new practices, alongside messages on the health benefits of improved S&H practices. Household visits should focus on one issue per visit, with a message specific to that issue supported by relevant IEC materials; messages

needed to be reinforced by more repeat visits to households, as bringing about behaviour change on S&H requires substantial follow-up.
- **Links to water supply**: it was emphasized that improvements in hygiene had to be better linked to investments in water supply to reduce problems linked to the unavailability or limitations of access to water, especially in the semi-arid Halaba Special *Woreda*.
- **Coordination**: it was thought that more collaboration between WASH stakeholders in the region was required, including active support from NGOs and donors to promote S&H under the government-led HEP.

In addition to measures to address the lack of resources for promotional activities – a key constraint to the BoH's ability to address each of the above issues – other forms of government action (such as by-laws) were seen as necessary to support the HEP. A number of the above issues will be addressed or supported through the *Kebele* WASH Teams (KWT) that will be established under the National WASH Implementation Framework (WIF) (see Chapter 1). Responsibilities of the KWT will include: to plan and manage the annual WASH inventory, to analyse the resulting data and take measures to improve coverage and sustainability of WASH services; to promote WASH; to support the WASH input of the water, health and agricultural extension workers, WASH Volunteers, and teachers; to support the input of community facilitators and other service providers when applicable; and to ensure sustainability of WASH services in collaboration with the *Woreda* WASH Team. KWTs are in the process of being set up in anticipation of bringing together the key government sectors and NGOs in a fully harmonized and integrated WASH Programme for planning, implementing, and reporting. Key principles underpinning the WIF and WASH delivery, including S&H, at local level are (1) integration, (2) harmonization, (3) alignment, and (4) partnership.

Promotion of S&H under health extension programmes: experiences in other countries

Early experiences

How does the Ethiopian approach to S&H promotion compare with experiences elsewhere, particularly in SSA, in delivering basic health services at relatively low cost? Encouraging community members to provide basic health services in their own communities is an approach dating back at least 50 years. Successful early examples in the 1950s include the 'barefoot doctors' in China and the village health volunteers in Thailand.[1] The inability of formal health services to deliver basic care to dispersed settlements led a number of developing countries to experiment with such approaches. Early examples in SSA include the village health worker (VHW) programmes carried out in Tanzania and Zimbabwe between the 1950s and 1970s, aimed at poverty eradication and rural development based on self-reliance (Lehmann and Sanders, 2007). The following outline of health extension initiatives in SSA is based primarily on (unless otherwise stated) the desk study conducted by Lehmann and Sanders (2007) – the most comprehensive survey to date and one that allows useful comparisons with the SNNPR experience (Gilson et al., 1989, had earlier provided an authoritative source).

Terminology

A range of terms are used to describe health providers recruited from communities including: basic health workers, outreach educators, rural health motivators, village health helpers, and of course community health promoters (CHPs). By their nature, community health workers (CHWs) – an umbrella term (WHO, 1990) – provide services to populations who are not reached by formal health services. In low-income countries with serious constraints on human resources for national health systems, the role assigned to CHWs is an important one, and includes reaching out to remote rural communities that are often marginalized and poor.

According to the widely recognized WHO definition, CHWs are: 'men and women chosen by the community, and trained to deal with the health problems of individuals and the community, and to work in close relationship with the health services. They should have had a level of primary education that enables them to read, write, and do simple mathematical calculations' (WHO, 1990: page number unavailable).

The functions that CHWs actually perform in different health systems vary substantially. A simple distinction is between generalist CHWs, such as those in Ethiopia, and specialist CHWs. The tendency for the past 20 years has been for more specialist CHW programmes focusing on maternal and child health, HIV/AIDS, TB care, malaria control, and treatment of acute respiratory infections. Food and water security, and sanitation and diarrhoea management, are considered secondary issues. The consequences of this choice of priorities are highlighted in the literature on S&H: 'While rarely discussed alongside the "big three" attention-seekers of the international public health community – HIV/AIDS, tuberculosis and malaria – one disease alone kills more young children each year than all three combined. It is diarrhoea, and the key to its control is hygiene, sanitation and water' (Bartram and Cairncross, 2010: 1).

Roles, priorities, and issues arising under health extension programmes

There has been a long and unresolved debate on just how many functions one generalist CHW can perform. Ultimately, the verdict will be given by the communities they serve. One study in Burkina Faso (Sauerborn et al., 1989) reported a problem of CHW credibility when two-thirds of ailments had to be referred on to the next level of care. For CHPs promoting S&H in SNNPR, the equivalent test of their effectiveness is in their ability to impart technical knowledge to households on how to improve the construction of latrines (to progress up the sanitation ladder), as well as the strength of their messaging on hygiene.

The evidence from experiences of health extension in a range of countries is that CHWs can improve access to, and coverage of, basic health services in communities. Too many large-scale programmes have failed in the past because of 'unrealistic expectations, poor initial planning, problems of sustainability, and the difficulties of maintaining quality' which have 'unnecessarily undermined and damaged the credibility of the CHW concept' (Lehmann and Sanders, 2007: v).

The essential features of well-performing CHW programmes are the appropriate selection of CHWs who then continue education/training together with proper supervision and support, backed by political leadership, and sustained by the substantial and consistent provision of resources (Lehmann and Sanders, 2007: vi) including equipment and materials for CHWs. Based on the findings of the RiPPLE research in SNNPR, these points have considerable resonance for the HEP in Ethiopia.

The issue of whether CHWs should be remunerated in some form, as an incentive to retain their services, is hotly debated. Experience suggests that CHW programmes operating on a completely voluntary basis encounter high attrition rates. Monetary incentives can increase retention, but 'often bring a host of problems, because the money may not be enough, may not be paid regularly, or may stop altogether' (Bhattacharyya et al., 2001). Non-monetary incentives, that make CHWs feel part of the health system, include supportive supervision and appropriate training, as well as relatively small things such as identification badges which can provide a sense of pride. Regular replenishment of supplies and materials also helps ensure that CHWs feel equipped to do their jobs (Bhattacharyya et al., 2001). Problems arise when some CHWs are paid, and some are not, and inconsistency between the practices of NGOs and government is liable to undermine the role of the CHPs under the HEP. The objective of the WIF is to effect full integration, alignment, and harmonization of service delivery approaches in a One WASH Programme. Depending on the extent of achievement of that objective, this inconsistency between NGO and government practices in Ethiopia should be reduced in future.

After the enthusiasm in the 1960s and 1970s for health extension initiatives, there was a decline following the economic recession in the 1980s that brought about reductions in public expenditure. CHW programmes in SSA suffered from World Bank-led structural adjustment, which emphasized private-sector health system development and reduced government spending on community health projects.

The HIV/AIDS crisis re-focused attention on CHWs for specialist roles, e.g. in South Africa (Blanchet et al., 2006). The evidence of CHW programmes from Gambia, Ghana, Madagascar, South Africa, Tanzania, and Zambia suggests that CHWs can enhance the performance of specialist community-level health programmes, particularly in the delivery of treatment for malaria, HIV/AIDS, and TB.

The CHWs in Rwanda were introduced in the late 1990s in response to the shortage of health staff after the death of many medical personnel during the civil war and genocide. They fulfil a generalist role in promoting S&H, including handwashing, use of latrines, boiling of water before drinking, and 'maintaining general cleanliness' (Rodriquez Pose and Samuels, 2011: 29; based on World Bank and Rwanda Ministry of Health, 2009 and Chambers, 2010). This is part of a broad role of sensitizing the population on a range of health issues as part of steps to decentralize health services. According to a recent study, the Rwandan CHWs have been integrated effectively into the health delivery system and the work of these volunteers – who live and work in their own communities – has taken health-care services into remote rural areas (Rodriquez Pose and Samuels, 2011). The sensitization of community members by the CHWs on health, hygiene, nutrition, and sanitation issues has helped increase people's awareness of the importance of taking care of their own health. CHW work on nutrition, although relatively recent, is already showing positive results, including a significant impact on child mortality (Rodriquez Pose and Samuels, 2011).

As for the transport constraints raised by the HEWs and CHPs in SNNPR, similar experiences can be seen elsewhere in SSA, with many health workers confined to clinics and health centres (Cairncross et al., 2010).

Conclusion

Prevention: the HEP in Ethiopia is an example of a system of primary health care with a strong preventive element, in contrast to health policies in many developing countries that focus predominantly on treatment by medication (Mara et al., 2010). Incorporating promotion of S&H, the HEP has received sustained political and institutional support from the Ethiopian Government through the MoH at federal level, as well as the regional BoH.

Rising numbers: the number of HEWs in Ethiopia continues to rise, with salaries revised upwards recently by the government to increase retention rates. In addition, the number of health posts in the *kebeles* is being extended (confidential interview with the representative of a leading NGO actor in the national WASH movement, 2009, personal communication).

Task force: The MoH leads the recently-established National Sanitation and Hygiene Task Force, which brings together government agencies, donors, and NGOs to support the HEP. Promotion of S&H under the HEP, CLTS, and CLTSH (including a strong focus on hygiene – handwashing, safe water collection and storage) has now been widely adopted by the Ethiopian Government.

Collective community process: CLTSH adds a community triggering process to the approach to promotion of S&H used by the SNNPR BoH post-2003 (as studied by RiPPLE) by bringing together the members of each community in a process that is collective, rather than via individual households. The CLTSH approach is compatible with the health extension system and the special triggering methods employed by CLTS are, as one key informant described it, blended into the existing promotion of S&H. The first task of the Hygiene and Sanitation Task Force has been to develop national guidelines on CLTSH. The CLTSH Training and Implementation Manual was completed recently and is ready for implementation. It is recognized that the awareness and desire for action facilitated by CLTSH requires intensive follow-up. Such capacity-building of human resources in health is one of the strategic objectives of the 2011–15 action plan for S&H, the Strategic Action Plan. There will be training in all regions to upgrade the skills of the HEWs, including the 'training of trainers' in CLTSH.

National WASH inventory: as noted above, the national WASH inventory is underway to ascertain the current status of access to water supply and sanitation in Ethiopia.

Motivation: the observations of the HEWs and CHPs consulted in the 2010 study in SNNPR (Behailu et al., 2010) were discussed with the BoH by the RiPPLE Regional Coordinator. The BoH is now looking at ways to strengthen the supervision and support provided to HEWs and CHPs and to provide a distinctive item of clothing (an apron or gown) to give CHPs a common visible identity – a potential motivator.

GLoWs Initiative: as for strengthening links between hygiene and water access, RiPPLE has supported the Guided Learning on Water Supply and Sanitation (GLoWS) initiative in its pilot project with a vocational training centre and *woreda* staff in SNNPR. GLoWS provides guided self-learning of small *woreda*-based teams to develop community WASH action plans covering water supply improvement, water quality risk management, sanitation and hygiene promotion, management of water points, and financial book-keeping. The approach leads to community action including agreements with *woreda* management on co-funding of interventions presented in the WASH action plans. This initiative could have the potential to strengthen the coordination of hygiene promotion interventions and water investments at local level.

Funds required: the lead responsibility for the promotion of basic social services such as S&H rests with government – the federal MoH and each regional BoH – including provision of funds. According to Bilal et al. (2011), an MDG needs assessment conducted in 2004 recommended scaling up the HEP and made a number of financial recommendations. It projected a requirement for additional funding over the five years from 2005 to 2010 from all sources (government and external), equivalent to US$3.48 per capita per year (peaking at $4.55 per capita in 2015). This was above the $7.13 average per capita total spend on health in Ethiopia in 2005, of which 28 per cent was borne by government. By 2008, the government had increased its contribution to total health spending by 77 per cent (Bilal et al., 2011). That still left a gap in available national resources for health, including for the HEP, as confirmed by the key informant interviews conducted by RiPPLE in 2011.

Supplementary finance: the health sector actors consulted (including an MoH official) emphasized that, for HEP-strengthening measures, more financial resources were needed than are available from government funds, including for salaries and better equipment and materials for HEWs/CHPs. Because of low sanitation coverage levels, total capital investments are in the range of $795 million per year ($692 million rural and $102 million urban) to meet the UAP target by 2015 – or roughly $10 per person (WSP, 2011). Following the government's policy of zero subsidy self-supply sanitation facilities, capital investments will be largely borne by households, while the government focuses on advocacy for appropriate low-cost technologies and promotion of S&H behaviour change. The anticipated public expenditure of around $30 million annually for rural sanitation will be mainly going towards institutional sanitation and promotion. Seven million dollars is anticipated to come from government sources and go primarily towards HEW salaries, while the remaining $23 million will come from donor sources (WSP, 2011). The government will, in other words, be depending on further external funding from development partners for S&H promotion and service delivery. Donors and NGOs should avoid setting up parallel HEPs that by-pass government, and target their contributions to fill the HEP resource gaps, following the lead of UNICEF and USAID – donors already providing direct support to the HEP. This should become easier in future with the WIF in place, which aims (as noted above) to achieve full integration, alignment, and harmonization of service delivery approaches under the One WASH Programme.

These supplementary resources for sanitation and hygiene promotion under the HEP are essential for sustained progress on promotion and achievement of the national targets of universal access to basic sanitation and 84 per cent access to improved sanitation by 2015, as well as reduction of infant and child mortality rates. Exemplary leadership has already been shown by the Ethiopian Government in setting the direction and ambition of its health extension effort.

Notes

1 'Barefoot doctors were health auxiliaries who began to emerge from the mid-1950s and became a nationwide programme from the mid-1960s, ensuring basic health care at the brigade (production unit) level. Partly in response to the successes of this movement and partly in response to the inability of conventional health services to deliver basic health care, a number of countries subsequently began to experiment with the village health worker concept' (Lehmann and Sanders, 2007: 5).

References

Bartram, J. and Cairncross, S. (2010) 'Hygiene, sanitation and water: forgotten foundations of health', *PLoS Med* 7(11): e1000367.
Behailu, S., Redaie, G., Mamo, D., Dimtse, D., and Newborne, P. (2010) 'Promoting sanitation and hygiene to rural households in SNNPR,

Ethiopia – experiences of health extension workers and community health promoters', *RiPPLE Working Paper 15*, Research-inspired Policy and Practice Learning in Ethiopia and the Nile Region (RiPPLE), Addis Ababa. All RiPPLE papers available from: <www.rippleethiopia.org/> [accessed July 2012].

Bhattacharyya, K., Winch, P., Leban, K., and Tien, M. (2001) *Community Health Worker Incentives and Disincentives: How they Affect Motivation, Retention and Sustainability*, Basic Support for Institutionalizing Child Survival Project (BASICS II), USAID, Arlington.

Bibby, S. and Knapp, A. (2007) 'From burden to communal responsibility – a sanitation success story from the Southern region of Ethiopia', *WSP Field Note*, Water and Sanitation Program (WSP) Africa, World Bank, Nairobi.

Bilal, N.K., Herbst, C.H., Zhao, F., Soucat, A., and Lemière, C. (2011) 'Health extension workers in Ethiopia: improved access and coverage for the rural poor', in P. Chuhan-Pole and M. Angwafo (eds.), *Yes Africa Can – Success Stories from a Dynamic Continent*, chapter 24, World Bank, Washington, D.C.

Blanchet, K., Keith, R., and Shackleton, P. (2006) 'One million more – mobilising the African diaspora healthcare professionals for capacity building in Africa', Conference Report, 21–22 March 2006, Save the Children, London.

Bureau of Health (BoH) (2006) *Ignition document No.8*, Southern Nations, Nationalities and People's Region (SNNPR) Bureau of Health, Hawassa.

Cairncross, S., Bartram, J., Cumming, O., and Brocklehurst, C. (2010) 'Hygiene, sanitation and water: what needs to be done?' *PLoS Med* 7(11): e1000365.

Chambers, V. (2010) 'Maternal health services in Nyanza and Kavumu Villages, Nyamagabe District, Rwanda', preliminary fieldwork findings, (unpublished).

Gilson, L., Walt, G., Heggenhougen, K., Owuor-Omondi, L., Perera, M., Ross, D., and Salazar, L. (1989) 'National community health worker programs: how can they be strengthened?' *Journal of Public Health Policy* 10: 518–32.

Haddis, A. (2008) 'The status of sanitation and hygiene promotion programmes in Ethiopia: situation analysis', *RiPPLE Synthesis Paper*, RiPPLE, Addis Ababa.

Lawn, J. E., Rohde, J., Rifkin, S., Were, M., Paul, V.K., and Chopra. M. (2008) 'Alma-Ata 30 years on: revolutionary, relevant, and time to revitalise', *Lancet* 372: 917–27.

Lehmann, U. and Sanders, D. (2007) *Community Health Workers: What do we Know about Them? The State of the Evidence on Programmes, Activities, Costs and Impact on Health Outcomes of Using Community Health Workers*, Department of Human Resources for Health, WHO, Geneva.

Mara, D., Lane, J., Scott, B., and Trouba, D. (2010) 'Sanitation and Health', *PLoS Med* 7(11): e100363.

Ministry of Health (MoH) (2005), *Health Sector Strategic Plan, Health Sector Development Programme HSDP-III, 2005/06–2009/10*, Federal Democratic Republic of Ethiopia (FDRE), Addis Ababa.

MoH (2007) *Health Extension Program in Ethiopia: Profile*, Health Extension and Education Center, MoH, FDRE, Addis Ababa.

MoH (2010) *Health Sector Development Programme IV, 2010/11–2014/15, Final Draft*, FDRE, Addis Ababa.
MoH (2011) *National Hygiene and Sanitation Strategic Action Plan for Ethiopia 2011–2015, Final Draft*, FDRE, Addis Ababa.
Newborne, P. and Samuels, F. (2010) 'Making the case for sanitation: opening doors within health', *ODI Background Note*, Overseas Development Institute (ODI), London.
Newborne, P. and Smet, J. (2008) 'Promotion of sanitation and hygiene to rural households in the Southern Region of Ethiopia', *RiPPLE Synthesis Paper*, RiPPLE, Addis Ababa.
Rehfuess, E., Bruce, N., and Bartram, J. (2009) 'More health for your buck: health sector functions to secure environmental health', *Bulletin of the World Health Organization 87*: 880–2, World Health Organization (WHO), Geneva.
Rodriquez Pose, R. and Samuels, F. (2011) 'Rwanda's performance in health: leadership, performance and insurance', *ODI Development Progress Story*, ODI, London.
Sauerborn, R., Nougtaara, A., and Diesfeld, H.J. (1989) 'Low utilization of community health workers: results from a household interview survey in Burkina Faso', *Soc Sci Med* 29(10): 1163–74.
Shiferaw, T. and Mariam, S. (2003) 'Assessment of health problems associated with water supply, sanitation and hygiene education in SNNPR', unpublished research paper.
Tefera, W. (2008) 'Technical issues of sanitation and hygiene in Mirab Abaya and Alaba: a case study report from the Southern Nations region of Ethiopia', *RiPPLE Working Paper 2*, RiPPLE, Addis Ababa.
Terefe, B. and Welle, K. (2008) 'Policy and institutional factors affecting formulation and implementation of sanitation and hygiene strategy – a case study from the Southern Nations Region (SNNPR) of Ethiopia, *RiPPLE Working Paper 1*, RiPPLE, Addis Ababa.
United Nations Children's Fund (UNICEF) (2004) 'What works for children in South Asia – community health workers', UNICEF Regional Office for South Asia, Kathmandu.
Vaughan, S. and Tronvol, K. (2003) 'The culture of power in contemporary Ethiopian political life', *African Studies Review* 48(2): 192–3.
WHO (1978) *Declaration of Alma-Ata* [online], WHO, Geneva. Available from: <www.euro.who.int/__data/assets/pdf_file/0009/113877/E93944.pdf> [accessed July 2012].
WHO (1990) *The Primary Health Care Worker: Working Guide*, WHO, Geneva.
WHO and UNICEF (2010) *Progress on Sanitation and Drinking Water: 2010 Update*, WHO, Geneva.
WHO and UNICEF (2012) *Progress on Sanitation and Drinking Water: 2012 Update*, WHO, Geneva.
World Bank (2005) *Ethiopia: A Country Status Report on Health and Poverty, Volume 1: Executive Summary, Report No.28963-ET*, World Bank, Washington D.C.
World Bank and Rwanda Ministry of Health (2009) *Rwanda: A Country Status Report on Health and Poverty*, Rwanda Ministry of Health and World Bank Africa Region, Kigali.

Water and Sanitation Programme (WSP) (2011) 'Water supply and sanitation in Ethiopia: turning finance into services for 2015 and beyond', *An AMCOW Country Status Overview*, WSP, Nairobi.

About the authors

Peter Newborne is a self-employed researcher and consultant specializing in water and sanitation, and water resources management, as well as environment, policy, and programmes. He is Research Associate to the Overseas Development Institute (ODI) Water Policy Programme. After reading humanities at the University of Oxford, he trained and worked for 10 years with firms in commercial law/contracting, based in London and Paris, acting for private and public clients in the banking, engineering, energy, real property, and other sectors. Further to a Master's in Development at the Ecole des Hautes Etudes en Sciences Sociales in Paris, he spent eight years with the World Wide Fund for Nature (WWF) contributing to sustainable development projects. In his current role, he conducts research studies and evaluations, and carries out consultancy assignments, on water and sanitation, and environment, in developing countries – for governments, NGOs, and industry representatives. Peter is a fluent French and Spanish speaker.

Anu Liisanantti works as a Programme Officer for ODI's Water Policy Programme. A post-graduate in human rights, she has previously worked in the editorial team for an HIV information charity, producing publications for people personally affected by HIV/AIDS as well as for medical professionals worldwide. She has also worked in the immigration/asylum sector, developing support services for people who find themselves in new surroundings.

CHAPTER 5

Sustainability of water services in Ethiopia

Nathaniel Mason, Alan MacDonald, Sobona Mtisi, Israel Deneke Haylamicheal and Habtamu Abebe

Ethiopia has made significant progress in extending access to improved water sources under its Universal Access Plan (UAP). Although data are contested, all sources confirm the strong upward trajectory. However, the ability of the country to sustain progress is difficult to predict. One key challenge is ensuring that investment translates into sustainable services that continue to meet users' needs in terms of water quantity, quality, ease of access, and reliability. Although data are limited, available evidence suggests that many schemes provide unreliable services or fail completely. Service sustainability is not a new issue in Ethiopia, or elsewhere in sub-Saharan Africa (SSA). The available evidence suggests that perhaps 40 per cent of handpumps are non-functional in SSA; in Ethiopia, official data suggest that 20–30 per cent of schemes have failed, or experience frequent outages. But a long-standing emphasis on capital investment and new infrastructure, coupled with weak monitoring and evaluation (M&E), has tended to obscure the problem, and few rigorous studies have been carried out on this topic. In this chapter, we review the evidence from Research-inspired Policy and Practice Learning in Ethiopia and the Nile Region (RiPPLE) research in two Ethiopian woredas *(districts) – Halaba Special* woredas *and Mirab Abaya – looking at water coverage, the number of non-functioning water schemes, and the factors that determine service sustainability, focusing particularly on rural water supply. Drawing on Ethiopian and wider regional research, we then highlight lessons and recommendations for addressing the problem at different decision-making levels.*

Introduction

Sustainable access to water supply is central to social and economic development, improving health and educational achievement, reducing child mortality, and improving livelihoods (Hutton and Haller, 2004). But these benefits are not sustained if access to water supply itself is not sustainable.

While there is very limited data available on water service sustainability, it has been estimated that in most developing countries, 30–60 per cent of rural water supply schemes are not functioning at any given time (Brikké and Bredero, 2003). In sub-Saharan Africa (SSA), the proportion of non-functional schemes has been estimated at almost 50 per cent, with most breaking down within three years of construction (ibid.).

Ethiopia has developed a plan to extend access to safe water. The ambitious UAP, launched in 2005, has been instrumental in galvanizing political and financial support for water supply and sanitation as a means of alleviating poverty (see Chapter 1), but sustaining services remains a huge challenge. Coverage data based on systems installed and assumed number of people served tell us little about the services people actually receive over time, as noted in Chapter 2. A recent high-level review of service delivery highlighted 'increased sustainability of infrastructure' as a key priority (AMCOW, 2011: 3).

Commentators have proposed different 'recipes' for sustainability, with community management high on the list of ingredients (see Chapter 3). In addition, factors such as gender sensitivity, partnership with local government and the private sector, and sufficient levels of cost recovery for basic maintenance and repair have also been emphasized (Brikké, 2002; Carter et al., 2010). In Ethiopia, the recent strategic shift by the government towards lower-cost technologies and 'facilitated self-supply' (see Chapter 3) is a response to the challenge of delivering and sustaining services in low income areas (AMCOW, 2011).

This chapter draws on RiPPLE research to look at the factors that affect the sustainability of water supply systems and services in Ethiopia, drawing on field work conducted in Halaba Special and Mirab Abaya *Woredas*. There are over 700 *woredas* in Ethiopia, so the research provides only partial insights. However, the chapter also draws on wider international experience to inform the discussion and conclusions.

Conceptual framework

What do we mean by sustainability? More specifically, the sustainability of what, and for whom?

In simple terms, sustainability is about: 'whether or not WASH services and good hygiene practices continue to work over time. No time limit is set on those continued services and accompanying behaviour changes. In other words, sustainability is about permanent beneficial change in WASH services and hygiene practices' (Carter et al., 2010: 2). In this chapter we use this definition, with a focus on water services, but draw a distinction between functionality and sustainability, and also between the service itself and the system used to provide it (Box 5.1).

Five key aspects can be separated to help understand the underlying drivers of service sustainability, highlighted in Figure 5.1. We argue that sustainability is more likely to be achieved when there is a balance between all five, represented by their intersection.

In brief, we can summarize as follows:

- Technical determinants, including the siting, design, and construction of water systems used to withdraw and deliver water to users.

Box 5.1 What is a water service?

A water service is sustainable if it continues to work over time, with service itself defined in terms of the quantity and quality of water accessible to users over time. Specific indicators include:

- Quantity, measured in litres per capita per day (lpcd).
- Quality, in terms of one or more separate indicators of chemical and biological quality.
- Distance from a household or centre of a community to a water point.
- Number of people sharing a source, often termed 'crowding'.
- Reliability, in terms of the proportion of the time the service functions to its prescribed level.

Monitoring the services accessed by individuals over time and space is clearly difficult. This is one reason why planners have focused on systems and the extension of new supplies, with assumptions then made about service levels using government standards (see Chapter 2) to determine water coverage.

Source: Moriarty et al., 2011; see also Butterworth et al., 2012

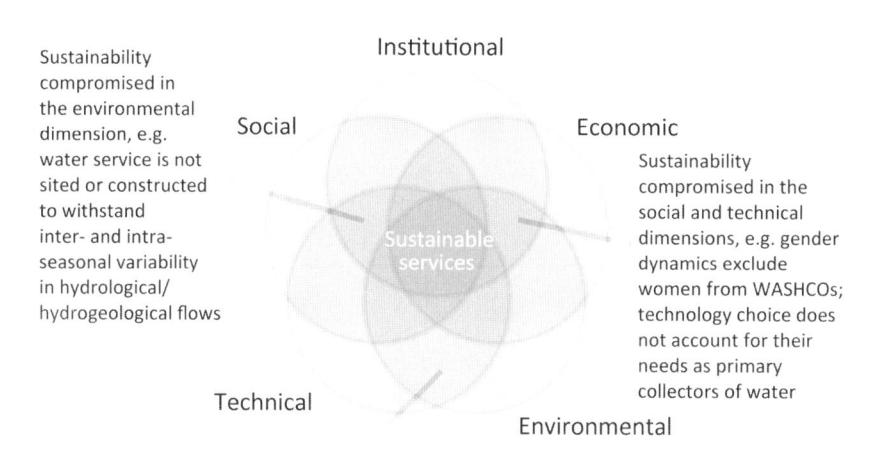

Sustainability compromised in the environmental dimension, e.g. water service is not sited or constructed to withstand inter- and intra-seasonal variability in hydrological/hydrogeological flows

Institutional

Social

Economic

Sustainable services

Technical

Environmental

Sustainability compromised in the social and technical dimensions, e.g. gender dynamics exclude women from WASHCOs; technology choice does not account for their needs as primary collectors of water

Sustainability compromised in the institutional, social and economic dimensions, e.g. lack of legal registration of WASHCOs inhibits ability to bank or borrow to undertake rehabilitation; WASHCO suffers loss of legitimacy in eyes of community; WASHCO finances further constrained by increasing non-payment for services by community

Figure 5.1 Conceptual framework for sustainability of water services

- Social determinants, including the relations and networks between individuals and communities.
- Institutional determinants: the formal and informal rules and structures governing the management of water supply schemes.
- Financial determinants: financial resources from various sources to meet all costs for long-term viability without undermining social development goals, such as poverty reduction.
- Environmental determinants, including the availability and quality (across time and space) of the water resource, linked to characteristics that affect the supply and its sustainability.

Water service sustainability – policy and practice

This section considers the progress of the water sector in Ethiopia on accelerating and sustaining access, highlighting the challenges to sustainability and emerging policy responses.

The substantial increase in resources and policy attention devoted to water, sanitation and hygiene (WASH) under the government's UAP has led to significant progress in extending access to safe water and sanitation. Addressing sustainability was a key aim of the original 2005 UAP, which aimed to do so in the first two years of implementation by focusing on rehabilitation and maintenance of existing schemes (MoWR, 2006a). The UAP reflected current global debates on sustainability, adopting such principles as demand-responsive approaches (DRAs), community contributions for operation and maintenance (O&M), and 'the participation of relevant bodies, especially women' (MoWR, 2006b: 6–7) in an effort to strengthen local ownership of water services and their sustainability (see Chapter 3).

While all sources confirm the positive trajectory in access, precise data are contested (Chapter 2). However, there is a general consensus that coverage estimates, based on the number of systems built and assumed number of people served from construction onwards, overestimate access to services. This highlights the difference between systems and services explained earlier, and the pitfalls of estimating coverage by counting the number of systems implemented without considering whether they are in fact providing the planned and desired level of service. The UAP's most recent revision estimated that close to 50,000 schemes were 'not functioning' for at least a few days each year – almost 30 per cent of the total (MoWE, 2010). However, even estimates from the United Nations Children's Fund (UNICEF) and the World Health Organization (WHO) Joint Monitoring Programme (JMP), derived from household surveys, provide only a partial snapshot of the problem as technology type is used as a crude proxy for the level and quality of service provided (Moriarty et al., 2011).

Research carried out for RiPPLE using the Water Economy for Livelihoods (WELS) framework highlights seasonal variation in functionality and service levels. While this is associated most obviously with climatic variation, it may be exacerbated by, for example, the coinciding of the peak labour period with the long dry season, with access compromised by long queues at water sources and the need to divert household labour to income-generating activities (Coulter et al., 2010).

Such seasonal challenges may intensify with climate change as rainfall becomes increasingly unpredictable (Chapter 7). The climate change challenge for sustainability should also be seen alongside expected population growth of nearly 90 per cent by 2050, which will place more pressure on existing services (Calow and MacDonald, 2009).

These observations highlight the need to go beyond conventional measures of coverage when considering service sustainability. The data challenge – knowing who has access to what services and where – will be partially addressed by the National WASH Inventory (NWI). As discussed in Chapter 2, however, this will capture functionality data for only a single point in time.

Ethiopia faces continued capacity and funding challenges at the local level. As is pointed out by Lockwood and Smits (2011) and in this book (see Chapter 1), government capacity remains very low. This problem is exacerbated by the decentralization of responsibilities but not finance, and the tendency for donor agencies and international non-governmental organizations (INGOs) to privilege capital investment in new systems (the hardware) rather than support structures and capacity for the maintenance and rehabilitation of existing schemes. The One WASH programme (see Chapter 1) instituted in 2006 has helped to consolidate sector efforts, although with greater impact on the implementation of new services than the maintenance and rehabilitation of existing ones.

The new draft UAP includes detailed human resource development plans, setting available capacity against required capacity by job type. Technical roles (handpump technicians, drillers, mechanical engineers) are the most difficult to fill (MoWE, 2010: 57) with serious implications for O&M, and there is a lack of clarity on how more than 120,000 skilled and professional staff are to be recruited, trained, and retained. One suggestion is to set up Operation and Maintenance Support Units (OMSUs) to support communities managing their own schemes, which will eventually evolve from public–private partnerships into full private entities. The new WASH Implementation Framework (the WIF – FDRE, 2011) also places the private sector at the centre of sustainability (McKim, 2011: 7; MoWE, 2010).

Water services sustainability in Ethiopia – learning from experience

This section draws on RiPPLE case studies conducted in Halaba Special and Mirab Abaya *Woredas* in the Southern Nations, Nationalities, and People's Region (SNNPR). The studies, carried out between November 2007 and February 2008, investigated the extent of, and reasons for, problems with service sustainability. Qualitative and quantitative methods, including water-point mapping; focus group discussions; and knowledge, attitude, and practice (KAP) surveys were used to trace the causal chains leading to unsustainable services. The two *woredas* have some similarities, including a prevalence of intestinal parasites and diarrhoeal disease, but differ significantly in the type of water supply systems used.

In Halaba the groundwater table requires deep boreholes, connected by distribution networks to water points. Water supply coverage had been estimated at around 40 per cent (BoFED, 2006), with 37 per cent of schemes estimated to be non-functional (AW-WRDO, 2007). The study found that only 24 of the 76 rural *kebeles* (communities) in the *woreda* had potable water supply from boreholes, distributing to a total of 65 water points. Ten schemes (42 per cent), and 40 water points (65 per cent) were non-functional – a greater water supply challenge than previous estimates suggested.

In Mirab Abaya the hydrogeology permits a range of technologies: hand- and machine-dug wells, boreholes, and protected springs. Thirty of the 70 schemes were found to be non-functional (43 per cent), of which 11 had been abandoned (Abebe and Deneke, 2008). Before the study, non-functionality was reported at 26 per cent (MAW-WRDO, 2007).

Scheme breakdowns were attributed to the technical failures of pumps and generators in most cases in Halaba. But these failures persisted for a range of social, institutional, and financial reasons, including the absence of follow-up support from the *woreda* or zone and insufficient training for operators. Environmental factors, such as water table drawdown and turbidity, were identified as the immediate causes of non-functionality for six of the schemes in Mirab Abaya. This ultimately raises questions as to whether these schemes were located and designed appropriately.

Technical challenges

Choice of technology can determine how easily sustainability can be ensured in relation to other aspects. The practical difficulties in involving communities in technology choice should not exclude them from planning, and technologies should be as user-friendly as possible. In Halaba Special *Woreda*, with its deep water table, WASHCOs and the *Woreda* Water Resource Development Office (WWRDO) preferred submersible, rather than mono-lift, pumps which were viewed as more prone to failure and requiring significant manpower to start (not always available).

Community participation from the outset can ensure sustainability by embedding an understanding of technology upkeep, maintenance and, proper usage. Despite being relatively simple, almost half of the hand-pumped wells in Mirab Abaya (20 of 48) were non-functional, with 'inappropriate use' reported as a major cause of failure by both WWRDOs and users. The lack of involvement of the primary users – women – in the planning and management of water services, can have done little to enhance familiarity and ownership of the technology. Research has shown the diverse benefits of such involvement for scheme functionality and women's empowerment (Fisher, 2006).

As noted previously, scheme functionality is one key element of service sustainability. The average round trip water collection time observed in Halaba Special *Woreda* was five hours. Such a heavy time burden may encourage users to revert to using unsafe but more local sources, especially during the wet seasons when surface water is more abundant, or to restrict water use. In both case studies consumption was generally below the 15 lpcd service level specified as 'adequate' in the UAP.

Social challenges

Involving communities effectively in the planning and management of their water services requires an understanding of socio-cultural norms – the attitudes and relationships that inform community interest in and usage of services.

Gender is key, given the time-consuming and physically demanding burden that insufficient, distant, and poor quality water supply places on women and girls – those typically responsible for collecting water and managing household water, sanitation, and hygiene. This makes it especially important to involve women in planning and managing the water services in which they have such a high stake.

The participation of women throughout the project cycle is emphasized in Ethiopia's sector policies, but the case studies suggest their continued exclusion. Focus group discussions with female users of water schemes indicated that WASHCOs rarely include women members; for some, the all-female focus group was the first chance to air their views (Box 5.2).

Observational evidence from schemes where women hold the majority of WASHCO positions – those implemented by the NGO Water Action in Halaba Special *Woreda* – indicated that they have better financial management and higher user satisfaction than those dominated by men. Various stakeholders confirmed that the model followed in Halaba Special *Woreda* was beneficial for scheme management, though time-intensive (Deneke and Abebe, 2008a), with women often unaware of their rights and opportunities to participate.

Box 5.2 Unheard voices

'No-one has ever before heard our voice with regard to water supply, which is women's major concern. Today, even though you are here not to provide us with water, we feel as if we have had a result. Because you are here at least to listen to what women say about water'. (elderly woman, Lower Lenda)

Source: Deneke and Abebe, 2008b

Regardless of gender, however, all water service management requires well-motivated personnel – often a question of social acceptance as WASHCO positions are voluntary (BoWR SNNPR, 2002). Where there are material incentives, however, these can distort: for example the availability of a *per diem* for WASHCO members to travel to the *woreda* office to report breakdowns encourages a culture of dependence even for relatively minor problems (Deneke and Abebe, 2008: 23–6).

These examples illustrate the challenge of designing criteria and procedures to overcome social relations and attitudes that may compromise service sustainability. These criteria and procedures comprise the institutional aspect considered in the next section.

Institutional challenges

WASHCOs are prominent institutional structures at the most decentralized level of communities and *kebeles*, but their impact depends on the existence of institutional rules and their effective implementation. Explicit rules for governing 'downward' accountability, from WASHCO to community, are limited to a stipulation that WASHCOs should report to communities every three months on income and expenditure. But this rarely happens, and a vague expectation that interaction between WASHCO and community will be formalized on an ad-hoc basis (BoWR SNNPR, 2002) often leaves users disenfranchised and unable to hold anyone to account for poor services (Deneke and Abebe, 2008: 41; Deneke and Abebe, 2008: 24).

WASHCOs lack formal legal status at present, so *woreda* finance offices will not audit them, creating a climate for weak financial management – cited as a reason for the replacement of some WASHCO members, and a source of dissatisfaction among communities (Deneke and Abebe, 2008). That said, informal arrangements may be sufficient in some cases: the lack of legal status has not stopped schemes opening bank accounts – predominantly with microfinance institutions in Mirab Abaya *Woreda*.

The WWRDO of Mirab Abaya attempts to visit WASHCOs to identify problems and respond to maintenance requests. However, with less than half of its positions staffed, few motor vehicles, and no budget for running costs, these activities are severely constrained – suggesting that the causal chain for sustainability failures can be traced from the institutional to the financial. In Halaba Special *Woreda* the WWRDO maintenance team lacks the equipment for major maintenance, and relies on three functional motorbikes to visit schemes, the furthest over 100 km away. However there are also significant gaps in skills and experience at *woreda* level in many cases, which limit the effectiveness of WWRDO planning, technical support, and monitoring.

While RiPPLE's sustainability case studies focused on the community up to *woreda* level, they also provide perspectives on capacity gaps at levels of government. The SNNPR Bureau of Water Resources (BoWR) has only one crane for major maintenance, for example. Delays attributable to lack of capacity, unclear communication channels, and inadequate motivation at all levels mean that obtaining support for major maintenance from the BoWR can take a minimum of three months in Halaba. In Mirab Abaya obtaining support can take up to a year as the Zonal Water Resource Development Office (ZWRDO) provides a level of support before the regional BoWR.

Given the failures of existing arrangements to provide systematic support to communities and WASHCOs in their management and maintenance of water services, it is not surprising that alternatives are being sought. The latest policy initiative to create OMSUs and increase the role of the private sector is an example, and there are positive experiences emerging from other countries in terms of the establishment of such units (e.g. India – Rajeev, 2012), and in information transmission between user-community and service agent through mobile and 'smart handpump' technologies (Hope, 2012; Rajeev, 2012). However, these tend to work best where there is a minimum density of water points that creates economies of scale for service agents. In rural Ethiopia, this may be difficult to achieve. More generally, new initiatives can increase institutional fragmentation and worsen coordination problems if not carefully crafted and piloted.

Financial challenges

Cost appears to lie behind many instances of unsustainable water services, with insufficient funds blamed for problems including lack of technical capacity and spare parts. As Cardone and Fonseca (2003) point out, cost recovery for a sustainable service requires all costs throughout the service lifetime to be met from different funding sources – users, government, and development partners. However, financial sustainability also requires that available funds are used effectively and raised equitably.

Cost-effective services require sound financial management to prevent the misuse of scarce funds. Only one of the schemes visited in Halaba Special *Woreda* had a coherent book-keeping system, and standard practice is for WASHCOs to collect revenue from the tap attendant on an ad-hoc basis. However, WASHCOs were trying to improve financial management by, for example, issuing receipts for water payments and in some cases banking savings. The good financial management practices of some schemes in Halaba Special *Woreda* indicate that it is possible to recover regular O&M costs from user fees, even where a low unit price (e.g. 10 or 15 cents per 25-litre can) is charged. However, whether this level of cost recovery is sufficient to fund major repair, or the upgrading and extension of services, is questionable. Certainly the international evidence, patchy though it is, suggests that the full costs of sustaining handpump services may be many times greater than the costs (of minor repairs) users are typically asked to meet (Baumann, 2006).

Raising funds equitably is vital to sustainable cost recovery and use of services: if poor users are priced out it deprives the service of revenue, and deprives users of safe water. While the capacity of WASHCOs to set complex tariff regimes is limited, an average tariff covering O&M costs should be affordable (Fonseca, 2003), though community consultation may be needed to identify those who may struggle (e.g. for cash contributions at certain times of year), and to find ways to cross-subsidize the poorest and most marginalized. On average, tariffs in Mirab Abaya were lower than in Halaba Special *Woreda*, and WASHCOs offered either a monthly fixed price or on-spot fee, with free water available to the poorest. In Halaba Special *Woreda*, however, the community perceived the tariff as too high to purchase water for everyday activities. Community consultation on tariff levels was rare and limited to male community members.

Just as at WASHCO level, where a *per diem* encouraged needless trips to seek *woreda* support, financial incentives can interact in unexpected ways at other administrative levels. In both case study *woredas*, the view was expressed that an expectation of supplementary NGO funding for water services constrained the amounts released by the *woreda*. Representation and dialogue are therefore critical in the politicized local government budgeting process: members of Mirab Abaya WWRDO felt that direct representation in the *woreda* cabinet would help ensure attention for water issues – a challenge also at zonal level. Allocations have fallen short of requests in recent years, but actual disbursements increased in the last year for which data are available.

Given the constraints, other solutions are being sought to increase the availability and effectiveness of finance, including facilitated self-supply, the Community-Managed Project (CMP) mechanism, and multiple-use water services (MUS – Chapter 3).

Self-supply is arguably an extension of DRA, with full responsibility for technology choice, financing, and implementation entrusted to the community or, more likely, the household. There are concerns that the new policy emphasis on self-supply may mean large-scale implementation precedes the development of vital support elements, such as finance, credit, and marketing capacity (Sutton, 2010). While households will have a clear incentive to maximize the sustainability of wells they have paid for and built themselves, poorly sited or constructed water services will fall into disuse or pose a health risk, just as in other implementation approaches. The provision of Water Extension Workers to provide backstopping support and guidance in *kebeles* with substantial self-supply activities (McKim, 2011: 65) will be critical to counter these risks.

The CMP approach is viewed as a scaled-up version of existing schemes based on the Community Development Fund (CDF) approach used by the Finnish–Ethiopian Rural Water Supply and Environmental Programme (RWSEP). Ownership of a CMP scheme would be entrusted to beneficiary communities from the outset, with communities receiving the responsibility and funds to plan and implement their own schemes, rather than relying on external agencies for implementation. The WIF foresees stringent criteria to determine community eligibility for the CMP mechanism. However, there is no dedicated programme to increase the limited number of communities that would currently be eligible. Initial monitoring from pilot regions suggests that the CMP approach is associated with high functionality rates, though these are short-term findings and as with all snapshot functionality figures, tell us nothing about service levels and whether these meet user needs.

MUS, detailed in Chapter 3, attempts to maximize the value extracted per 'drop'. RiPPLE case study evidence from Boro Gutu *Woreda* in Oromia Region suggests that multiple-use services (irrigation and domestic water) have higher up-front costs and may be more complex to manage, but exhibit better overall cost–benefit ratios than single-use (irrigation or domestic only) schemes (Adank, et al., 2008). MUS is expected to enhance productive uses that could increase communities' ability to afford maintenance and repair (Faal et al., 2009).

A final critical component is a functioning market around spare parts and skills. The emphasis on increased private-sector involvement in the revised UAP acknowledges the need for proper incentives to consolidate spare part and pump supply across a region to achieve sufficient scale (MoWE, 2010: 36). However, neither the UAP nor the WIF provide details on capacitating the private sector beyond the evolution of OMSUs and their transition to private entities.

The financial aspect is not necessarily the end of the causal chain, however. Hydrogeological conditions may determine operational and repair costs by, for example, mandating more costly technology, especially where motorized pumping is required. Such environmental aspects are considered in the next section.

Environmental challenges

The environmental aspect of water services sustainability brings us back to the resource itself. Most rural water services rely on groundwater, which is not invulnerable to degradation, but provides a natural buffer against climate variability and drought – responding much more slowly to meteorological conditions than surface water – and generally requires little treatment (Calow et al., 2010).

Wells and boreholes are less likely to be seasonally dry if they are carefully located in good aquifers with enough porosity and permeability for storage and movement of groundwater (MacDonald and Calow, 2009) – see Figure 5.2. Hydrogeological maps may be supplemented with geophysical techniques to identify likely groundwater resources and the amount of investment required to develop a sustainable water point. During drilling or construction, pumping tests and close control over the process can enhance prospects for sustainability.

The quality of the resource also has an obvious bearing on the quality of the water service. Two naturally occurring contaminants, fluoride and arsenic, are particular health concerns (Hunter et al., 2010). RiPPLE has identified that over ten million people could be at risk of fluorosis in Ethiopia,

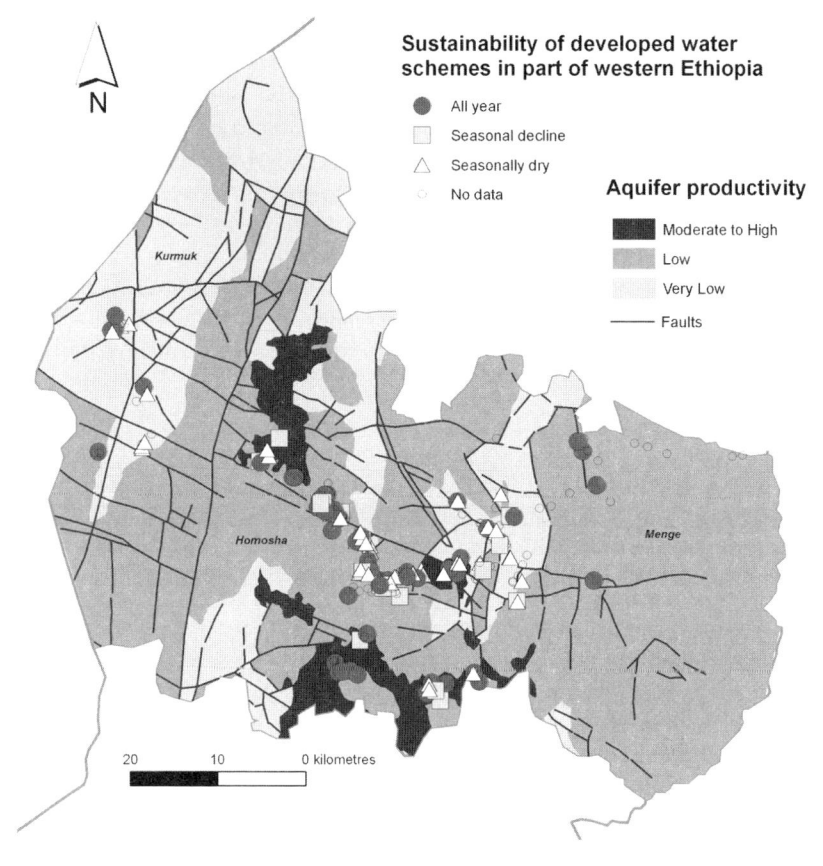

Figure 5.2 The sustainability of water points related to aquifer productivity for three *woredas* in Benishangul-Gumuz

Source: MacDonald et al., 2009

though wells or springs that are only a short distance apart may have radically different fluoride concentrations. Groundwater quality (and potentially long-term health and productivity) can also be compromised by surface contaminants, including animal and human excreta. Promotion efforts such as Community-led Total Sanitation (CLTS) have rapidly reduced open defecation, but not confined excreta (MoH, 2011). Visible quality problems such as turbidity can prevent people using even technically safe sources, as observed for two schemes in Mirab Abaya.

While falling water tables are often reported as major concerns, data on water tables across Africa are limited. Calow et al. (1997; 2010) demonstrated that in low permeability aquifers, immediate drawdown due to pumping has the greatest effect on water levels, which can be reduced by siting wells and boreholes in more productive parts of an aquifer. Constructing wells or boreholes to levels well below the dry season water table and testing them accordingly can enhance sustainability, given natural seasonal changes of several metres in the water table.

Longer-term changes in water-tables are harder to measure and predict, with a complicated relationship between rainfall and recharge that is mediated through land use and other factors. However, studies indicate that rainfall above 500 mm per annum will *generally* provide sufficient recharge for rural water supplies (Edmunds, 2008; Calow and MacDonald, 2009). The impacts of climate change on groundwater availability and quality are uncertain, and much depends on the timing, frequency, and distribution of rainfall events – still difficult to model – rather than long-term average trends (Box 5.3). Developing water supplies that can accommodate current natural variation will help ensure resilience to future climate change (Howard et al., 2010; Bonsor et al., 2010; Calow et al., 2011).

Box 5.3 Implications of climate change for water services sustainability

Global warming will lead to higher rates of evapotranspiration and a likely increase in the intensity and variability of rainfall (Christensen et al., 2007; Conway, 2011), and most scientists agree that both surface run-off and groundwater recharge will become less reliable. In Ethiopia specifically, annual rainfall is actually forecast to increase in highland areas.

The potential impacts of climate change on water services include (after MacDonald et al., 2009; Bonsor et al., 2010; Howard et al. 2010; Calow et al., 2011; MacDonald et al.,2011):

- Unimproved, shallow water sources are likely to be more vulnerable to increased climate variability because sustainability is closely coupled to rainfall.
- Improved rural water sources that access groundwater over 20 m below ground surface are likely to be more sustainable, however, a significant minority of people could be affected by more frequent and longer droughts – particularly in areas with limited groundwater storage (Figure 5.3).
- Water supplies reliant on groundwater close to the coast are at increased risk of salinization.
- Extreme weather events such as storms and floods will lead to a greater destruction of water infrastructure, from large city supplies to small community supplies, and increase the risk of contamination.
- Some water supply technologies will have a higher degree of resilience to climate change, strengthening the rationale for using multiple sources throughout a year, each with a different risk profile.
- Access to water rather than absolute water availability will remain the key determinant of water security in most areas.
- An additional complicating factor is the impact of climate change on demand. Abstraction of reliable groundwater for non-domestic purposes such as irrigation could increase, though this could in turn enhance water security by strengthening livelihoods and ability to contribute to maintenance and repairs.

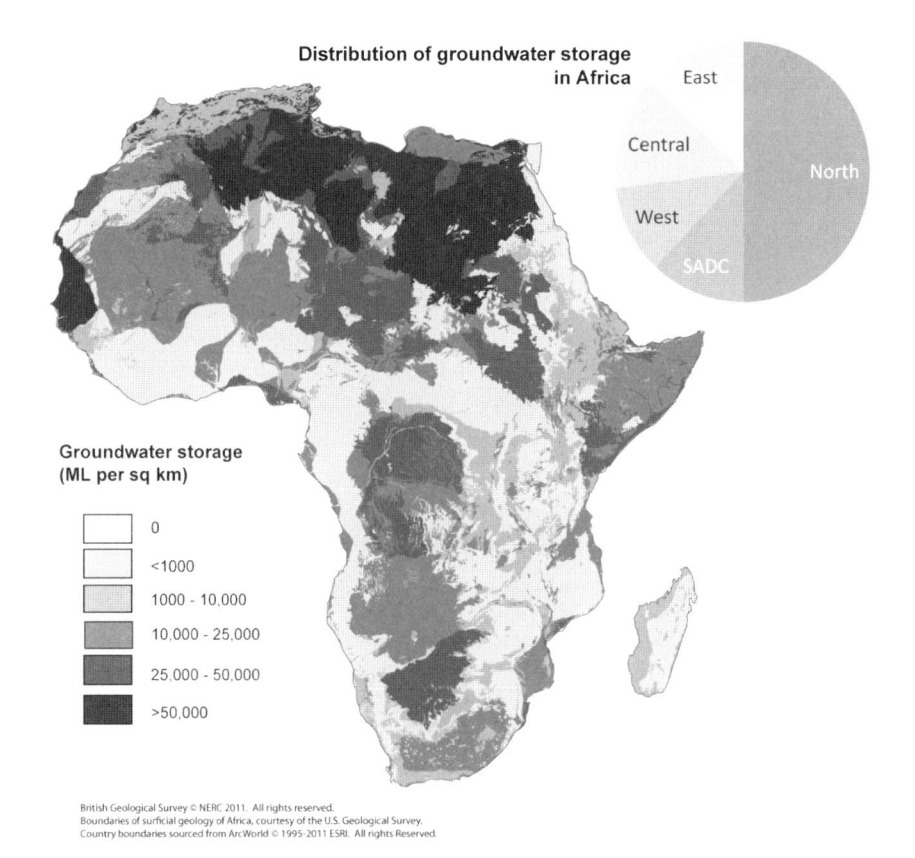

Figure 5.3 Estimated groundwater storage in Africa

Note: The large aquifers in North Africa contain a significant proportion of Africa's groundwater, but are 'fossil' aquifers because they do not receive contemporary recharge from rainfall. Less productive aquifers throughout much of SSA have less water, but storage is still sufficient to support domestic and minor productive uses. High average annual recharge will increase the resilience to short-term (i.e. interannual) climate variability.

Source: British Geological Survey © NERC 2011 in MacDonald et al., 2012

Though the environmental aspect is considered last in the list of five used to frame this discussion, it is not necessarily the end of the causal chain for unsustainable services. In many instances, choice of technology, social sensitivity, robust institutions, and a realistic approach to long-term economics can do much to mitigate the innate risks associated with the water resource itself.

Conclusion

The sustainability of water systems and services in Ethiopia depends upon a complex interaction of technical, social, financial, institutional, and environmental factors. Efforts to extend and sustain water services will founder without a clearer understanding of these contributory factors, and their influence on both systems and services.

The following recommendations draw on the RiPPLE case studies and wider learning, with a focus on Ethiopia but with broader applicability to rural water services in SSA.

- Monitoring and evaluation should go beyond new schemes and increased coverage. The National WASH Inventory (NWI) is a positive step, with its focus on functionality if not broader service levels, but effectiveness will depend on: access to data; capacity for its use in planning and budgeting; and the regularity of the process. The implicit incentive structure arising from targets for increased coverage also needs to change; for example targets could focus on sustainability as well as the number of new schemes. In the short term, the effects of seasonality should be built into monitoring by capitalizing on existing information resources such as seasonal WASH assessments around food security – with seasonality integrated into the NWI in the longer term.
- Capacity building is required at all levels, but especially among WWRDOs and WASHCOs. This includes technical training for scheme maintenance and operation, but also training on the broader institutional skill set including planning, budgeting, and monitoring. The UAP ambition to recruit and train over 120,000 skilled professional staff requires significant resourcing and careful planning aimed at training – and retaining – new recruits. Vocational training, including the innovative Guided Learning on Water Supply and Sanitation (GLoWS) programme currently being piloted through vocational colleges (Chapter 8) – can help meet this goal.
- The revised UAP and associated WIF emphasize the role of the private sector in supply chains, in O&M, and in scheme implementation. Private-sector capacity remains limited, however, in part because of high entry barriers and public sector monopolies, and because profits are likely to be thin or non-existent when dealing with dispersed rural communities. Nonetheless, support for the establishment of OMSUs is an encouraging step, even if transitions to public–private or full private status cannot be achieved where water point densities are low.
- Water service sustainability depends on sufficient financial resources and effective financial management. The preceding recommendations require additional funds, but those funds need to be used effectively, and bottlenecks that limit the absorptive capacity (Chapter 1) of local government need to be addressed. WASHCOs need particular support

to develop better financial management skills and systems, and to raise the participation of women. However, the ability of user-groups to fund all ongoing operation and maintenance, including major repairs, is questionable. Cost-sharing arrangements between communities, government, civil society organizations (CSOs), and donors may therefore need to be extended beyond project planning and implementation phases.

- Alternative systems for rural water supply, such as MUS and self-supply, have potential benefits for service sustainability. MUS fits well within a broad concept of sustainable services – incorporating the idea of different sources for different uses, in different seasons – since it demands a more holistic assessment of water needs, matched to available resources. The resource base could be enhanced by development of local, decentralized water storage as a buffer against variability. In the case of self-supply, greater understanding is needed of how communities and households can be better supported to ensure sustainability of their own services. There is a particular need to better understand where, and for whom, self-supply is appropriate, and to clarify the role of local government in implementation and backstopping.
- From a resource perspective, groundwater development provides an opportunity to extend reliable water services, at reasonable cost, to dispersed rural populations. A key advantage of groundwater is the buffer aquifer storage provides against rainfall variability – now and in future. To make the most of this potential, however, water systems (MUS, self-supply, shallow and deep boreholes, etc.) need to be closely matched to hydrogeological conditions as well as existing and potential user demand.

References

Abebe, H. and Deneke, I. (2008) 'The sustainability of water supply schemes: a case study in Mirab Abaya woreda', *RiPPLE Working Paper 4*, RiPPLE, Addis Ababa.

Abebe, H. and Deneke, I. (2008) 'The sustainability of water supply schemes: a case study in Alaba Special woreda', *RiPPLE Working Paper 5*, RiPPLE, Addis Ababa.

Adank, M., Jeths, M., Belete, B., Chaka, S., Lema, Z., Tamiru, D., et al. (2008) 'The costs and benefits of multiple uses of water: the case of Gorugutu *Woreda* of East Hararghe Zone, Oromiya Regional States, eastern Ethiopia', *RiPPLE Working Paper 7*, Research-inspired Policy and Practice Learning in Ethiopia and the Nile region (RiPPLE), Addis Ababa. All RiPPLE papers available from: <www.rippleethiopia.org/> [accessed July 2012].

African Ministers' Council on Water (AMCOW) (2011) 'Water supply and sanitation in Ethiopia: turning finance into services for 2015 and beyond' *AMCOW Country Status Overview*, Water and Sanitation Programme (WSP), Nairobi.

Alaba Woreda Water Resources Development Office (AW-WRDO) (2007) *Alaba Special Woreda Water Resources Development Office Schemes Assessment Report,* AW-WRDO, Halaba.

Baumann, E. (2006) 'Do operation and maintenance pay?', *Waterlines,* 25(1): 10–12 <http://dx.doi.org/10.3362/0262-8104.2006.033>.

Bonsor, H. C., MacDonald, A. M., and Calow, R. C. (2010) 'Potential impact of climate change on improved and unimproved water supplies in Africa', *RSC Issues in Environmental Science and Technology,* 31: 25–50 <http://dx.doi.org/10.1039/9781849732253-00025>.

Brikké, F. (2002) *Operation and Maintenance of Rural Water Supply and Sanitation Systems: A Training Package for Managers and Planners,* World Health Organization (WHO), Geneva.

Brikké, F. and Bredero, M. (2003) 'Linking technology choice with operation and maintenance in the context of community water supply and sanitation', *Reference Document for Planners and Project Staff,* WHO and IRC International Water and Sanitation Centre (IRC), Geneva.

Bureau of Finance and Economic Development (BoFED) (2006) *Terminal Evaluation Report on Water Action, Alaba, and Surrounding Water and Environment Development Programme, 2001–2005,* BoFED, Hawassa.

Bureau of Water Resources (BoWR) (2002) *SNNPR Rural Water Supply Implementation Plan,* BoWR, Hawassa.

Butterworth, J., Welle, K., Bostoen, K., and Schaefer, F. (2013) 'WASH sector monitoring', in R. Calow, E. Ludi and J. Tucker (eds) *Achieving Water Security: Lessons from Research in Water Supply, Sanitation and Hygiene in Ethiopia,* pp. 49–68, Practical Action Publishing, Rugby, UK.

Calow, R. and MacDonald, A. (2009) 'What will climate change mean for groundwater supply in Africa', *ODI Background Note,* Overseas Development Institute (ODI), London.

Calow, R.C., Robins, N.S., MacDonald, A.M., Macdonald, D.M.J., Gibbs, B.R., Orpen, W.R.G., Mtembezeka, P., Andrews, A.J., and Appiah, S.O. (1997) 'Groundwater management in drought prone areas of Africa', *International Journal of Water Resources Development,* 13: 241–61 <http://dx.doi.org/10.1080/07900629749863>.

Calow, R.C., MacDonald, A.M., Nicol, A.L., and Robins, N.S. (2010) 'Ground water security and drought in Africa: linking availability, access and demand', *Ground Water,* 48(2): 246–56 <http://dx.doi.org/10.1111/j.1745-6584.2009.00558.x>.

Calow, R.C., Bonsor, H., Jones, L., O'Meally, S., MacDonald, A., and Kaur, N. (2011) 'Climate change, water resources and WASH: a scoping study', *ODI Working Paper 337,* ODI, London.

Cardone, R. and Fonseca, C. (2003) 'Financing and cost recovery', *IRC Thematic Overview Papers,* IRC, Delft.

Carter, R., Harvey, E. and Casey, V. (2010) 'User financing of rural handpump water services', paper presented at the IRC Symposium 2010 Pumps, Pipes and Promises, IRC, The Hague.

Christensen, J. H., Hewitson, B., Busuioc, A., Chen, A., Gao, X., Held, I., Jones, R., Kolli, R.K., Kwon, W.-T., Laprise, R., Magaña Rueda, V., Mearns, L., Menéndez, C.G., Räisänen, J., Rinke, A., Sarr, A., and Whetton, P. (2007) 'Regional climate projections', in S. Solomon, D. Qin, M. Manning,

Z. Chen, M. Marquis, K.B. Averyt, M. Tignor, and H.L. Miller, *Climate Change 2007: The Physical Science Basis. Contribution of Working Group I to the Fourth Assessment Report of the Intergovernmental Panel on Climate Change*, Cambridge University Press, Cambridge.

Conway, D. (2011) 'Adapting climate research for development in Africa', *WIREs Climate Change*, 2: 428–50 <http://dx.doi.org/10.1002/wcc.115>.

Coulter, L., Kebede, S., and Zeleke, B. (2010) 'Climate change futures of water: impacts on highlands and lowlands', *RiPPLE Working Paper 16*, RiPPLE, Addis Ababa.

Edmunds, W. M. (2008) 'Groundwater in Africa – Palaeowater, climate change and modern recharge', in S. Adelana and A. MacDonald, *Applied Groundwater Research in Africa, IAH Selected Papers in Hydrogeology 13*, Taylor and Francis, Amsterdam.

Faal, J., Nicol, A., and Tucker, J. (2009) *Mutiple-use water services (MUS): cost-effective water investments to reduce poverty and address all the MDGs*, RiPPLE and Multiple-Use Water Services Group, Addis Ababa.

Federal Democratic Republic of Ethiopia (FDRE) (2011) *The WASH Implementation Framework (WIF), Version July 2011*, FDRE, Addis Ababa.

Fisher, J. (2006) *For Her, It's the Big Issue. Putting Women at the Centre of Water Supply, Sanitation and Hygiene*, Water Supply and Sanitation Collaborative Council, Geneva.

Fonseca, C. (2003) 'Cost recovery: taking into account the poorest and systems sustainability', *Watershed Management for Water Supply Systems: Proceedings of the American Water Resources Association 2003 International Congress, New York, June 29–2 July 2003*, American Water Resources Association, Middleburg.

Hope, R. (2012) 'Smart handpumps and rural water security risk', *Proceedings of the International Water Security Conference, 16–18 April 2012*, University of Oxford, Oxford.

Howard, G., Charles, K., Pond, K., Brookshaw, A., Hossain, R., and Bartram, J. (2010) 'Securing 2020 vision for 2030: climate change and ensuring resilience in water and sanitation services', *Journal of Water and Climate Change* 1(1): 2–16.

Hunter, P. R., MacDonald, A. M., and Carter, R. C. (2010) 'Water supply and health', *PLoS Med* 7(11): e1000361 <http://dx.doi.org/10.1371/journal.pmed.1000361>.

Hutton, G. and Haller, L. (2004) *Evaluation of the Costs and Benefits of Water and Sanitation Improvements at the Global Level*, WHO, Geneva.

Lockwood, H. and Smits, S. (2011) *Triple-S, Water Services that Last. Lessons for Rural Water Supply: Moving towards a Service Delivery Approach. A Multi-country Synthesis*, IRC, The Hague.

MacDonald, A.M., Bonsor, H.C., Calow, R.C., Taylor, R.G., Lapworth, D.J., Maurice, L., Tucker, J., and Dochartaigh, B. E. (2011) 'Groundwater resilience to climate change in Africa', *British Geological Survey Open Report*, OR/11/031: 25pp, British Geological Survey, Edinburgh.

MacDonald, A. M., Ó Dochartaigh, B. É. and Welle, K. (2009) 'Mapping for water supply and sanitation (WSS) in Ethiopia', *RiPPLE Working Paper 11*, RiPPLE, Addis Ababa.

MacDonald, A.M. and Calow, R. C., (2009) 'Developing groundwater for secure rural water supplies in Africa', *Desalination*, 248: 546–56 <http://dx.doi.org/10.1016/j.desal.2008.05.100>.

Mirab Abaya Woreda Water Resources Development Office (MAW-WRDO) (2007) *Mirab Abaya Woreda Water Resources Development Office Schemes Assessment Report,* FDRE, Ethiopia.

McKim, C. R. (2011) *Draft National WaSH Program Implementation Framework,* Ministry of Water and Energy (MoWE), FDRE, Addis Ababa.

Ministry of Health (MoH) (2011) *National Hygiene and Sanitation Strategic Action Plan for Ethiopia Final Draft,* FDRE, Addis Ababa.

Moriarty, P., Batchelor, C., Fonseca, C., Klutse, A., Naafs, A., Nyarko, K., Pezon, C., Potter, A., Reddy, R., and Snehalatha, M. (2011) 'Ladders for assessing and costing water service delivery', *WASHCost Working Paper 2*, IRC, The Hague.

Ministry of Water and Energy (MoWE) (2010) *Revision of UAP (Water): Draft Report,* MoWE, FDRE, Addis Ababa.

Ministry of Water Resources (MoWR) (2006a) *Universal Access Program for Water Supply and Sanitation Services, 2006–2012. Part I – Rural* (originally published in Amharic, 2005), FDRE, Addis Ababa.

MoWR (2006b) *Universal Access Program for Water Supply and Sanitation Services. Part III: Strategy* (originally published in Amharic, 2005), FDRE, Addis Ababa.

Rajeev, K. J. (2012) 'WaterAid's mobile-enhanced hand pump maintenance innovations in rural India', *Proceedings of the International Water Security Conference, 16–18 April 2012*, University of Oxford, Oxford.

Sutton, S. (2010) 'Accelerating self supply: a case study from Ethiopia 2010', *Field Note No. 2010,* Rural Water Supply Network, St. Gallen.

About the authors

Nathaniel Mason is a Research Officer in the Overseas Development Institute's (ODI's) Water Policy Programme, where he works on financial and institutional aspects of water and sanitation services, as well as water security issues. He has previously worked as a consultant for the Water and Sanitation Programme (WSP) in Nairobi, on the African Ministerial Conference on Water (AMCOW) Country Status Overviews, and for WaterAid's policy team in the UK and Nepal.

Alan MacDonald is a principal hydrogeologist at the British Geological Survey where he divides his work between UK and international groundwater issues. He has published more than 40 scientific papers and is author of the book *Developing Groundwater: A Guide to Rural Water Supply* and editor of *Applied Groundwater Studies in Africa*. Much of his work focuses on the science base for sustainable development and management of groundwater, with a particular interest in climate resilience and poverty reduction. He has a BSc in geophysics from Edinburgh University and an MSc and PhD in hydrogeology from University College London.

Sobona Mtisi is a Research Officer in ODI, Water Policy Programme, with a background in sociology and demography, and holds a PhD in international development from the University of Manchester. Sobona has over seven years' research experience on water governance and policy processes and their implications for social and economic development in southern Africa.

Israel Deneke Haylamicheal is a lecturer at the Department of Chemistry, Hawassa University, and a freelance consultant. He has a BSc in chemistry, an MSc in environmental sciences, and is currently studying environmental technology and engineering in Europe. Mr Haylamicheal along with his colleagues has published research and review papers on obsolete pesticides and healthcare waste management and rural water supply. His research interests include water supply and sanitation, chemicals and hazardous waste management, and emerging contaminants and wastewater treatment.

Habtamu Abebe is a Policy and Research Officer at the Resource Center for Civil Society Groups Association (RCCSGA) based in Hawassa, Ethiopia. He holds a BSc in chemistry from Bahir Dar University and an MSc degree in environmental science from Addis Ababa University. Mr Habtamu has a background working on peace and conflict analysis, particularly in pastoralist and agro-pastoralist regions, and was heavily involved in the RiPPLE research conducted in the SNNPR region of Ethiopia. He has authored a number of papers.

CHAPTER 6

Water for livelihood resilience, food security, and poverty reduction

Josephine Tucker, Zelalem Lema and Samson Eshetu Lemma

Investments in water are central to Ethiopia's growth and climate adaptation strategies, to overcome the vulnerability to rainfall variations of millions of farmers and the national economy, and reduce the costs of poor access to water, sanitation and hygiene (WASH). Research Research-inspired Policy and Practice Learning in Ethiopia and the Nile Region (RiPPLE) programme has 'ground-truthed' international findings on the benefits of these investments, confirming that improved domestic water access is associated with reduced poverty, increased employment, and lower incidences of water-related disease. Small-scale irrigation – a key policy plank – enhances productivity, increases income, and mitigates drought risk at the household level, but more transformative economic impacts will require vastly improved infrastructure and markets. Improved livestock water provision could increase the important contribution of livestock to the agriculture sector, and enhance rangeland condition and pastoral livelihoods. Overall, embedding much-needed infrastructure development in basin management and watershed protection strategies will be necessary to avoid worsening degradation and sustain the benefits from investments.

Introduction

Securing access to water for human consumption, hygiene, and productive uses is central to tackling poverty and food insecurity and increasing resilience across sub-Saharan Africa (SSA). An inability to access enough water without spending excessive time or money or facing personal risk, and the unreliability of water for crops and livestock, helps to keep millions in poverty. This is rarely because of absolute water shortages, at least for domestic needs; it is the result of inadequate investment to ensure reliable access for all.

Poor water access undermines food security (food availability, access, and absorption) via three principal routes at the household level (Tucker and Yirgu, 2011; Calow et al., 2010):

128　ACHIEVING WATER SECURITY

1. Insufficient or poor quality water for domestic use (particularly hygiene) causes water-related disease, which in turn reduces the body's absorption of nutrients and increases susceptibility to other illness.
2. Long collection times (due to distance or queuing) reduce time available for work or education, principally for women and girls.
3. Insufficient water for livestock and crops (or other productive activities) limits food production and income generation.

These are interconnected in various ways (Figure 6.1).

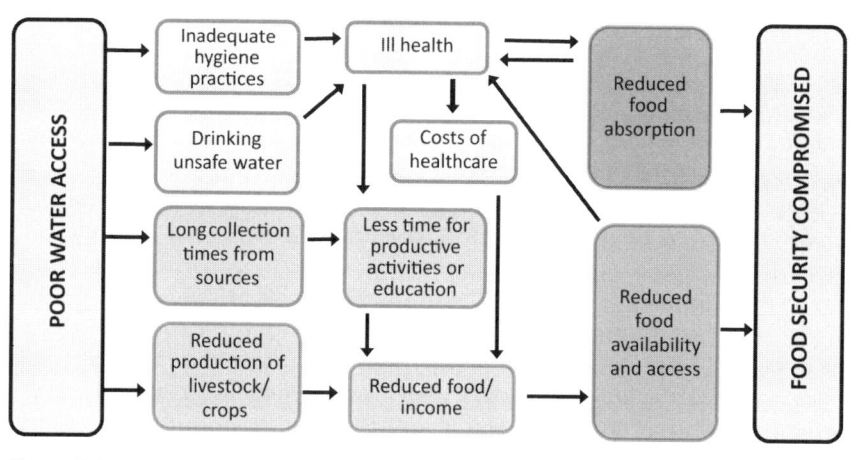

Figure 6.1 Causal pathways linking lack of water with food insecurity

Source: adapted from Tucker and Yirgu, 2011

As well as bringing benefits at the household level, improvements in access to domestic and productive water access at scale are necessary for growth and development (SIWI, 2005). The potential contribution of water to growth is high in Ethiopia, where rainfall is highly variable, water storage is minimal, and both livelihoods and gross domestic product (GDP) are vulnerable to hydrological variation because of dependence on rainfed agriculture and, at national level, hydropower. The World Bank (2006) estimates that this 'unmitigated variability' costs Ethiopia more than one-third of its growth potential and reduces GDP by up to 10 per cent in severe droughts. It also contributes to severe chronic food insecurity for millions. Escaping this vulnerability requires a minimum underlying 'platform' of water infrastructure for storage and distribution (Grey and Sadoff, 2007), the development of which must be accompanied by water resource management arrangements which ensure equity and sustainability. Realizing the economic benefits of water investments will also require simultaneous investment to overcome other major constraints to growth and productivity.

The following sections discuss available evidence on the role of water in livelihood security and wealth generation, with a focus on field-level evidence from Ethiopia. Firstly the focus is on domestic water, and then on agriculture.

Domestic water

Global analyses

Global studies attribute almost 2.5 million child deaths annually, and around six per cent of the worldwide disease burden in terms of disability-adjusted life years (DALYs),[1] to inadequate access to water supply and sanitation and poor hygiene practices (Prüss-Üstün et al., 2008; Bartram and Cairncross, 2010). Diarrhoea, and subsequent malnutrition, accounts for most of this disease burden.

It is estimated that universal access to basic improved water and sanitation facilities would save households US$565 million and national health budgets $11 billion in treatment costs, through prevention of infectious diarrhoea alone (Hutton and Haller, 2004). These are underestimates of the full benefits to be derived from universal access, as they do not include other illnesses which may result from poor water and sanitation access (e.g. water-washed diseases), secondary infections occurring following bouts of dirrahoea, or the long-term cognitive and physical impairment caused by chronic childhood illness and malnutrition (Guerrant et al., 2002; SIWI, 2005; Hunter et al., 2010; Bartram and Cairncross, 2010).

Bartram and Cairncross (2010: 3) provide detailed discussion of the linkages between WASH, health, and nutrition, concluding that 'with the possible exception of malaria and HIV/AIDS in Africa, it is hard to think of another health problem [than inadequate WASH] so prejudicial to household and national economic development'.

Time losses due to reliance on distant or unreliable services may be at least as costly. Estimates suggest that even meeting the water and sanitation Millennium Development Goal (MDG) targets would save around 40 billion working days worldwide,[2] accounting for 80 per cent of the potential total economic benefit of improved services (Hutton et al., 2007). Overall, global cost–benefit analyses find that improvements in WASH should yield $5–46 for every $1 invested, with the highest returns in least-developed areas (ibid.).

Evidence from Ethiopia

Global analyses of the impact of water supply on livelihoods and food security require ground-truthing, yet there is a dearth of detailed country evidence. RiPPLE has helped to fill this gap in Ethiopia. Indications are that benefits should be huge: early research found that people in some areas spend nine hours per day collecting water from unimproved sources in the dry season (Abebe and Deneke, 2008), while the World Health Organization (WHO) estimated that unsafe water and sanitation cause around 112,000 deaths and the loss of over 3.5 million DALYs per year (WHO, 2009).

RiPPLE conducted macro-level analysis using data from national Welfare Monitoring Surveys (Anderson and Hagos, 2008) to explore links between access to water and sanitation and several factors related to growth and poverty reduction: engagement in productive employment, school enrolment of children, health status, and the self-reported changes in households' food situation and overall welfare.

Data showed that a general increase had occurred in access to protected water sources and improved sanitation facilities between 1999/2000 and 2004/5. School enrolment and health status had also increased in most regions, and productive employment had slightly increased nationally, although with considerable regional variation.

Regression analyses using the 2004/5 data found that the main drivers of improvements in welfare and food situation experienced over the previous year (controlling for shocks such as deaths) were land ownership and rainfall shortages. The type of water or sanitation facilities did not have a significant effect. However, reported *improvements* in access to water and sanitation over the preceding year were significantly associated with improvements in households' (self-reported) food situation and overall welfare. Improvements in sanitation were not found to have an effect. Interestingly, no corresponding negative relationship was found between deterioration in water access and worsening of households' food and welfare situation.

Detailed analysis at *woreda* (district) level suggests that decreasing distance to water sources has a significant positive effect on school enrolment and employment of both men and women; the results vary depending on whether wet or dry season access is considered, and between regression techniques, but there are indications of a positive relationship. Overall this macro-level analysis was limited by the data available (no panel data were available for 1999/2000 and 2004/5, and many variables were scored categorically with few categories), but nonetheless it provides some evidence that improving water access brings economic benefits for households.

This was followed by a survey of 1,500 rural households in eastern Ethiopia (East Hararghe Zone, Oromia Region), and analysis of the connections between access to water, health, labour, and poverty (Hagos et al., 2008). The following discussion is based on this research and some further analysis of the data by the authors. Households with access to improved water supply were less likely to be poor (defined as having consumption expenditure below an inflation-adjusted poverty line of Ethiopian birr 1,821.05 (US$101), based on the 1995/96 official poverty line of ETB 1,075 or $60). They were more likely to be able to meet the minimum costs of food than those without (consumption expenditure above ETB 1,096.02 or $61,[3] a similarly adjusted food poverty line based on the official 1995/6 price of a basic food basket) (see Figure 6.2). At the time of the survey, 87 per cent of surveyed households without access to improved water supply lived below the poverty line, compared with around 67 per cent of those with access.

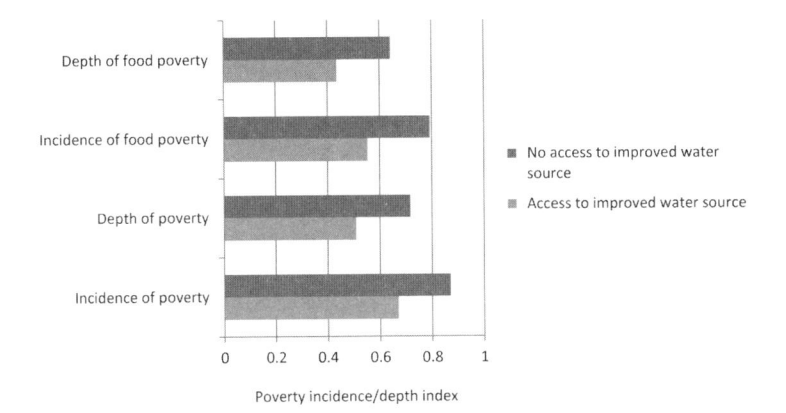

Figure 6.2 Incidence and depth of poverty among households with and without improved water supply in East Hararghe

Note: Incidence of (food) poverty = share of the population whose income is below the (food) poverty line; depth of (food) poverty = the gap between household income and the (food) poverty line

Source: Tucker, 2009, adapted from Hagos et al., 2008

Further analysis explored the routes by which access to improved water supply might reduce poverty. The first is via health improvements (see Figure 6.1). Hagos et al. (2008) found that over 30 per cent of households reported experiencing water-related disease in the previous year, including almost 15 per cent experiencing diarrhoea and 6 per cent malaria. There were also reports of skin conditions, which are considered water-washed as they are linked to poor hygiene. 4.4 per cent of individuals across all the surveyed households reported experiencing a waterborne or water-washed disease in the previous 12 months, and 3.3 per cent diarrhoea or dysentery. Highly significant differences are seen between age groups ($p < 0.0001$ for Pearson chi-squared test), with under 5s and over 65s most likely to have experienced these diseases (Table 6.1).

Table 6.1 Experience of water-related disease among different age groups

Age group (years)	Individuals experiencing waterborne or water-washed disease in the year before survey (%)	Individuals experiencing diarrhoea or dysentery in the year before survey (%)
0–4	5.8[1]	4.4[1]
5–15	3.5[1]	3.0[1]
16–64	4.3	3.5
>65	10.4[1]	6.6[1]

[1] Result significant at the 0.05 level, for Pearson chi-squared test

Source: author's own, using RiPPLE's household survey data

Focusing on diarrhoeal disease (including dysentery), the transmission of which is most directly related to WASH, further analysis found that access to improved water supply is associated with a significantly lower proportion of individuals reporting these diseases in the past year (2.9 per cent of those with access had experienced diarrhoea, compared with 3.8 per cent of those without). The effects of handwashing and sanitation require more investigation, but handwashing *with soap* after toilet visits seems to be associated with even greater, highly significant, reductions in rates of diarrhoea; just 1.9 per cent of those who report handwashing with soap had experienced diarrhoea. This is consistent with a systematic review which concluded that handwashing with soap could reduce the risk of diarrhoea by over 40 per cent (Curtis and Cairncross, 2003). A combination of improved water access and handwashing with soap seems to be particularly powerful: only 1.2 per cent of people with this combination of safe water and good hygiene practices had experienced diarrhoea. Data suggest that handwashing without soap has no effect, and in combination with lack of access to improved water supply, it may in fact increase the risk of diarrhoea. The reasons are not clear but the authors speculate that handwashing without soap may create a 'false sense of security' and encourage riskier behaviour when it comes to transmission of diarrhoeal disease (e.g. touching food). Full results from the above analysis are presented in Table 6.2. The use of a latrine also appears to have no effect, perhaps due to the poor hygienic condition of latrines or inconsistent use.

Table 6.2 Experience of diarrhoeal disease among individuals with differing access to improved water supply and handwashing practices

Improved drinking water supply	No access			Access		
Handwashing practices (after toilet visits)	None	With water only	With soap	None	With water only	With soap
Individuals having experienced diarrhoeal disease in the year prior to survey (%)	3.4	4.1	2.5[1]	3.5	2.7	1.2[1]

1 Result significant at the 0.05 level, for Pearson chi-squared test
Source: author's own, using RiPPLE's household survey data

Among those who sought treatment for these illnesses, health-care costs (including transport, consultation and medication) ranged from nothing (free consultation only) to over ETB 2,800 ($150), averaging ETB 131 (around $7.20) per person. Where medication was purchased this generally accounted for most of the total spend, averaging ETB 110 ($6) but reaching ETB 2,600 (just under $145) in one case. Given that 75 per cent of surveyed households (in fact a subset of just under 1,000 households for which data were available) have total annual expenditure of less than ETB 820 ($45), these costs are significant. If we assume the same disease incidence, treatment seeking rates and costs of transport and health care nationwide, across a population estimated at 82.9 million (World Bank, 2006), Ethiopian households are spending in the order of ETB 470 million (around $26 million) per year on treating these diseases. Households in which at least one working age individual (16–64) experienced a waterborne or water-washed disease lost on average ETB 96 (about $5.50) in income from working days missed, up to ETB 1,300 ($72) in some cases.

The second route is via time savings. Households with access to improved water were 14 per cent more likely to undertake off-farm employment than those without, and off-farm employment was found to significantly decrease household poverty (Hagos et al., 2008). This suggests that the reduced distances to water associated with access to an improved source (a relationship confirmed by the survey) may reduce poverty by improving households' ability to take up employment opportunities. Off-farm employment is also significantly increased by access to credit and possession of non-farm skills.

Such strong effects of water access, even in areas with high unemployment, indicate that poor access to water constrains people's ability to undertake off-farm employment. This is relevant for rural job creation programmes, in particular public works schemes targeting the most vulnerable, principally the Productive Safety Net Programme (PSNP). This is echoed by research in Bale Pastoral Region, which found that households sometimes sacrifice up to 20 per cent of their PSNP working days to collect water (Coulter, 2008). One method used by RiPPLE to understand these tradeoffs is a seasonal calendar of water access (see Figure 6.3).

The final route is through productive use of water. Many households reported collecting water for non-household purposes, often from multiple sources, but detailed data were only collected for irrigation (discussed below).

Agricultural water investments

Agriculture employs 65 per cent of people in SSA and contributes on average 34 per cent of GDP (Bach and Pinstrup-Andersen, 2008; UNECA, 2011). Agricultural growth is highly pro-poor (Ligon and Sadoulet, 2007), but is hindered by inadequate or unreliable access to water for production, and vulnerability to water-related hazards.

Water productivity in SSA is generally low because of unreliable water supply, poor market access (for produce and improved inputs), and the severe poverty and vulnerability that prevent small farmers from adopting risky but potentially high-return strategies (World Bank, 2008; Kemp-Benedict et al., 2011). Per capita food production in SSA has declined over the last 30 years, unlike other regions (Giordano, 2006), and remains vulnerable to climate variability and change as over 90 per cent of production is rainfed (McCartney and Smakhtin, 2010).

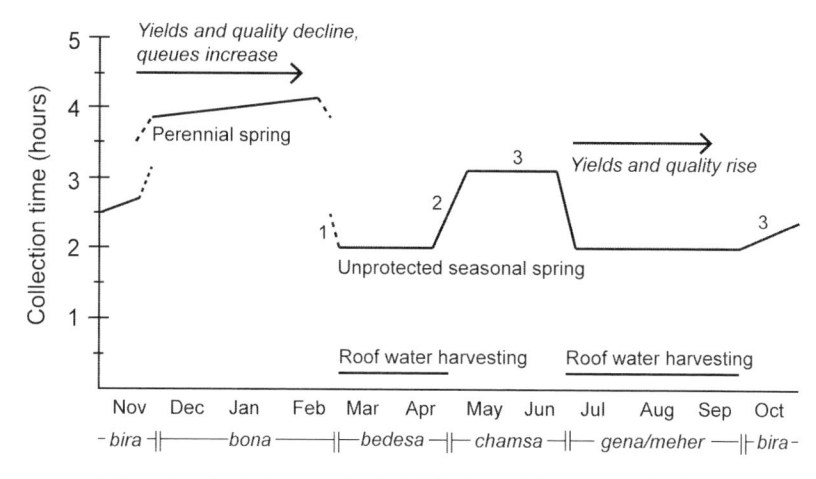

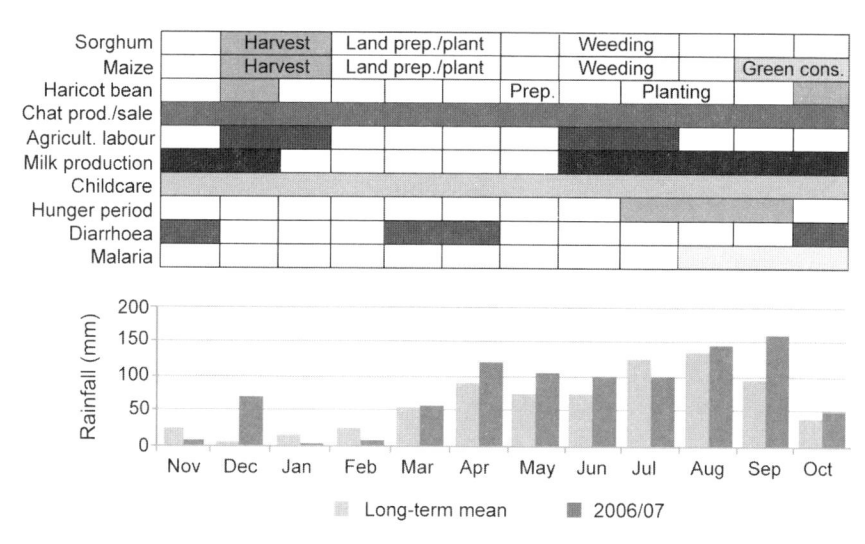

Figure 6.3 Seasonal calendars of water access

Note: Seasonal calendars can be used to shed light on periods of poor water access and their links with health and labour shortages. These calendars, showing a midland agricultural zone in eastern Ethiopia, clearly show the coincidence of high collection times in December to February with high demand for agricultural labour on which many poor households depend. The co-occurrence of diarrhoea and the use of unprotected, but convenient, springs in March and April, is also evident.

Source: Coulter et al., 2010

Ethiopia has challenging hydrology, yet water development and management are at a low level; in 2003 Ethiopia had just 43 m^3 of major surface water storage per capita, compared with over 6,000 in the USA (World Bank, 2006). Rainfall is intensely seasonal and variable, with frequent droughts and floods, while farmers mostly subsist on small rainfed plots, cropping once per year, with few improved inputs and no irrigation or water storage (Awulachew et al., 2007). Rain failures and rainfall at the wrong time frequently destroy crops, and water for livestock is often insufficient in dry periods.

Ethiopia's 2011 Growth and Transformation Plan (GTP) casts commercialization of smallholder agriculture as an engine of growth, to be supported by investment in small-scale irrigation, transfer of experiences from 'model farmers' via the extension system, and strengthening of markets and rural infrastructure. The GTP recognizes that a shift to higher-value production depends on improved water use, and ultimately on more sustainable resource management, and therefore promotes soil and water conservation and forest protection (MoFED, 2010). Much of Ethiopia is highly degraded with huge rates of topsoil erosion, and declining productivity combined with population growth drives cultivation of ever more marginal lands. Reversing the decline of ecosystem services provided by soils, forests and wetlands will be vital for sustainable agricultural growth (Abebe and Geheb, 2003).

Ethiopia's recent Green Economy Strategy (FDRE, 2011) includes raising crop and livestock productivity as one of four core strategies. Agricultural intensification is planned through extension of small-, medium- and large-scale irrigation as well as increased use of improved inputs and better residue management. Watershed management is also emphasized. Agricultural intensification is intended to reduce deforestation (although empirical evidence on whether intensification leads to land sparing is mixed – see Perfecto and Vandermeer, 2010). Ethiopia is increasingly seeking foreign direct investment to expand irrigation; according to Negash (2012), this is a response to a lack of domestic investment.

Small-scale irrigation is also a priority intervention in Ethiopia's National Adaptation Programme of Action (see Chapter 7). Because of its policy prominence, and high interest among regional and local stakeholders, small-scale irrigation was chosen as the subject of a RiPPLE case study.

Irrigation development for poverty reduction

International and Ethiopian experiences

In Asia, irrigation expansion bolstered growth and poverty reduction through direct income gains and stimulation of the rural economy (Hussain and Hanjra, 2004). The prospects for such spill-over effects are poorer in Africa because of structural constraints on rural growth and weak multipliers in rural areas: poor infrastructure; thin markets; and highly dispersed, poor populations (Lipton and Litchfield, 2003; Masiyandima and Giordano, 2007;

Eshetu et al., 2010). Irrigation development itself is hindered by weak institutions for irrigation and water resources management, and by high levels of poverty and poorly functioning markets that do not provide the means or incentive for household investment in irrigation, except where surface water can be easily diverted. Access to energy for pumping is very limited.

Less than five per cent (200,000 ha of a potential 3.7 million ha) of Ethiopia's irrigable land was developed in 2006 (World Bank, 2006), although more recent estimates go up to 18 per cent, with irrigation contributing around nine per cent of agricultural GDP (Hagos et al., 2009). Ethiopia is in the top six African countries in terms of irrigation potential from surface water sources (defined as the area irrigable from surface water where irrigation is predicted to yield a positive internal rate of return on investment), but without significant market improvements, investments in irrigation are unlikely to be highly profitable (You et al., 2010). Combined biophysical and socio-economic analysis indicates that the conversion of large areas to small-scale irrigation (here defined as diversion rather than dam-based, and irrigating smallholdings rather than commercial farms) would be profitable only where travel times to urban centres are under five hours (ibid.). This would exclude swathes of rural Ethiopia.

Nevertheless, irrigation allows farmers to increase yields, diversify into cash crops, and mitigate the risk of rain failure (Peden et al., 2007). In Ethiopia, the International Fund for Agricultural Development (IFAD) reported yield increases of 25–40 per cent from improved small-scale irrigation (up to 100 per cent from spring-based systems), with subsequent increases in farmers' financial and physical assets and food security (IFAD, 2009). Irrigation can also benefit nearby non-irrigators through increased food production and labour opportunities (Van Den Berg and Ruben, 2006), but mixed empirical findings on the equity impacts of irrigation point to the need for strong institutions to ensure inclusion of poor households, non-irrigators, and 'tail-end' users in wealth generation. Experiences suggest, though, that establishing effective institutions for irrigation management is difficult and requires long-term post-construction support (IFAD, 2009).

Scaling out irrigation will depend on available water resources. Recent research suggests that while there is plentiful groundwater across Africa to meet domestic needs, the development of large groundwater-based irrigation areas is unlikely to be sustainable in all regions (MacDonald et al., 2011; You et al., 2010). Measures to address potential upstream–downstream competition for water will be an essential complement to any investment in scaling out irrigation, as will natural resource management and soil water conservation programmes to maximize in-field water availability and maintain soil fertility and other ecosystem services under more intensive production.

RiPPLE research findings

RiPPLE's household survey (see above) found that access to irrigation is associated with significantly lower incidence, depth, and severity of both poverty and food poverty (Hagos et al., 2008). However the effect of irrigation is much smaller than the effect of improved access to drinking water.

Another RiPPLE case study examined the prospects for small-scale irrigation to enhance household resilience and generate pro-poor growth in eastern Ethiopia (East Hararghe Zone, Oromia Region). Selected communities had been irrigating for at least five years (spring-based schemes with water storage, serving between 120 and 1,300 households), in three agro-ecological zones: highland *dega*, midland *weyna dega* and lowland *kolla*. The following discussion of findings is based on Eshetu et al. (2010).

Irrigators reported an average 20 per cent increase in income, achieved mainly by producing high-value vegetables, increasing cash crop production to two or three crops per year and, in the lowlands, irrigating livestock fodder. This is far below the average 219 per cent increase in income per ha reported by Hagos et al. (2009) for irrigated smallholder systems over rainfed systems, and somewhat below increases of 25–100 per cent reported by IFAD (2009). This may be because water access was limited for some households, or because of differences in access to markets or other assets (rural finance, agricultural inputs or land). However, over 90 per cent of irrigating households surveyed reported increased income and better nutrition, and communities have reported new levels of wealth. Poorer households felt that their status had improved, and domestic conflicts reduced.

However, while some saw their income increase four-fold, others reported only small gains (Box 6.1). Income gains depend on access to household and community assets: land, labour, finance for inputs, and water supply. Poorer households often lack capital and labour to intensify production and have small landholdings; many reported that irrigation helped them to meet basic needs but they remained food-insecure and reliant on the PSNP. Female-headed households reported the least benefit, mainly due to labour constraints. Differences also relate partly to attitudes to risk; medium and poor households tended to use irrigation to reduce risks rather than to maximize income. The link between irrigation benefits and starting wealth is confirmed by RiPPLE research in Southern Nations, Nationalities, and People's Region (SNNPR – Abebe et al., 2010), as well as experience elsewhere in Ethiopia (Gebreselassie, 2010), although the latter finds that farmer decisions around land allocation to irrigation are also a critical determinant of impact and may decrease with overall landholding size. Differences in returns on irrigation are also attributed to land fertility, motivation, skills, and irrigation water availability. The latter relates partly to land accessibility and distance from the source, but corruption in water distribution was also widely reported (preferential treatment of friends and relatives, or of *chat*-producers over vegetable-growers).

Finally, all farmers face high costs accessing markets and receive low prices for produce. Inadequate storage facilities and inability to access large market centres prevent high-volume sales, while poor market information and monopolies by middlemen in small local markets restrict farmers' price-setting power.

There is limited evidence of positive spill-overs from irrigation. Some non-irrigating households reported expansion of trading activities, and greater opportunities to borrow food or cash within the community at times of scarcity, but overall, non-irrigators reported minor impacts. High levels of reinvestment by irrigating households in production are a more positive sign; the average reported annual spend on agricultural inputs has more than doubled, and on livestock inputs increased by almost half, mainly among relatively better-off households. Some households have also reinvested in trading and beverage production, although these remain small-scale and localized. Others have invested in children's education, with likely long-term benefits.

These findings confirm international experience suggesting that small-scale irrigation can substantially increase rural incomes and food security, but only as part of an integrated approach combining post-construction support for management institutions, land husbandry, access to inputs and finance, and investment in market development as well as attention to opportunities for female-headed households and non-irrigators.

Water for livestock

Livestock and water in sub-Saharan Africa

Livestock production is the principal livelihood activity of 20 million pastoralists in SSA. Livestock are also vital for agricultural livelihoods, providing draught power, manure for fertilizer or fuel, and milk and meat for consumption, as well as an important form of insurance and wealth accumulation (Opio, 2009). Eighty per cent of Ethiopian farmers plough with animals, and recent estimates value livestock's contribution at up to 45 per cent of Ethiopia's agricultural GDP, nearly a third of this from draught power (Behnke, 2010).

Peden et al. (2007) characterize livestock production in SSA as a 'high-potential opportunity to reduce poverty, increase productivity, boost investment returns on agricultural water development, and improve environmental sustainability'. However, effective policies to support livestock production are rare (Opio, 2009). Livestock are widely neglected in agricultural water planning and management (Peden et al., 2007), despite their significant water demands (mainly for fodder production), their impact on water resources and land, and the fact that water deprivation and long journeys to water, reduce milk production and overall productivity. RiPPLE research found that livestock in agricultural and agro-pastoral areas often consume insufficient water in the dry season, with likely impacts on their health, productivity and market value (Coulter et al., 2010; Tucker et al., forthcoming).

Water interventions in pastoral areas have often failed to meet needs or consider how water access governs human and livestock distribution. Water availability is a significant determinant of migratory patterns, and the location and spacing of water points is critical to prevent overgrazing and conflict. Permanent water points have often been developed in areas

Box 6.1 Contrasting experiences of irrigation

Mohamed Abrahim Abdulahi is considered 'better-off' in his community. He has 0.25 ha of land and three adult children who work the farm. He has quadrupled his income using motorized irrigation to grow vegetables, fruits, *chat* and coffee for market. He pays for water (ETB 100 ($6) per irrigation session) and other inputs (up to ETB 100 ($6) per year) and makes around ETB 4,500 ($250) per year. His family is now food-secure.

Credit: Addisu Delalegn

Asha Abdo lives in the same *woreda* but is considered 'poor'. She lost her husband and supports six children, only one of whom is working age, on 0.13 ha of land. Asha grows maize and sorghum for home consumption and irrigated *chat* for market, earning about ETB 900 ($50) per year but spending ETB 750 ($41) on labour, irrigation water, and inputs. She cannot afford to produce *chat* twice per year due to labour costs, and has only slightly increased her income using irrigation, although she considers her family's situation to be better than before irrigation was introduced.

Credit: Estifanos Worku

Source: Eshetu et al., 2010

previously used only for wet season grazing, allowing year-round use which results in pasture degradation, conflict, and spread of human and livestock disease (Gomes, 2006).

In some cases this has been followed by enclosure of grazing land and appropriation of water sources by particular groups, and the development of water points has sometimes been used deliberately to promote sedentarization (Nassef, 2012). According to Wilson (2007), while large amounts of money have been spent on water development in rangelands in recent decades, impacts on quality of life and drought survival have been minimal.

Pastoral water development in Ethiopia: synthesis of experience

RiPPLE, with Save the Children USA and Care Ethiopia, carried out a review of Ethiopia's water policies and interventions in pastoral areas (Nassef, 2012), on which the following section is based.

Pastoral water development in Ethiopia has exhibited many of the above problems. Top-down interventions, such as construction of permanent ponds, have encouraged permanent settlement (often intentionally) and led to overgrazing, erosion, and spread of disease. There are some recent signs of a shift in thinking, with more emphasis on grassroots participation to help planners understand seasonal patterns of resource use.

The World Bank- and IFAD-supported Pastoral Community Development Project (PCDP) recommends small temporary catchments to provide water in the wet season while preventing permanent settlement, alongside rehabilitation of existing water points and maintenance of access to rivers in dry season grazing areas. PCDP guidelines recommend that water points are 20 km apart and do not exceed a size which waters 4,500 cattle per day. The increased attention to multiple-use water services (MUS) in national policy is also positive, as in pastoral areas, humans and livestock use the same water sources. Finally, an adapted PSNP for pastoral areas has been piloted, which takes account of traditional management structures, seasonal movements, and community priorities, using participatory techniques.

However, the focus on new infrastructure promoting sedentarization, such as boreholes, continues, and some government programmes have overridden and weakened the customary pastoralist institutions which evolved to manage resources sustainably (see also Wilson, 2007). Development programmes in pastoral areas are also dwarfed by short-term emergency relief, under which water points are often hastily planned and constructed. The Ethiopian Government recognizes the need to move from emergency response to vulnerability reduction, and effective approaches for pastoral areas are urgently needed (see also Tucker and Yirgu, 2011).

Small positive shifts must be seen in the wider context of land use transformation in pastoral areas. Irrigation is expanding – often limiting dry season access to vital permanent water sources – and policies promote settled agropastoral livelihoods over mobile pastoralism. Sedentarization is encouraged in current Ethiopian policy (see for example FDRE, 2011: 71). Pastoral lowlands are also targeted for resettlement to reduce population pressure in the highlands. Many pastoralists might welcome new livelihood opportunities, but current approaches are often more prescriptive than empowering. Some large-scale land use transformation plans such as the Oromia Growth Corridor, however, include retention of large areas of rangeland and provide for multiple livelihood strategies, while policy emphasizes that sedentarization – although encouraged – must be voluntary (MoFED, 2010).

Conclusion

RiPPLE has generated ground-level evidence from Ethiopia on the vital role water security plays in human and economic development. Secure water for domestic and productive use underpins health, livelihood security, and economic opportunity. Ambitious programmes in the sector offer an opportunity for growth, poverty reduction, and the improvement of quality of life for millions; to maximize and sustain these benefits, and avoid unforeseen negative impacts, a number of lessons should be noted.

Domestic water

Ethiopia's drive to achieve universal access is impressive, but realization of the potential human and economic benefits requires a simultaneous focus on ensuring reliability of services (see Chapter 5) and on promoting good hygiene practices. Improved access to water supply is likely to enhance take up of new economic opportunities created by the government's investment programmes in rural areas. Continued poor access, however, may undermine their impact.

Water for agriculture

RiPPLE research confirms that small-scale irrigation – a policy priority – improves income and food security for users, but that economic spill-overs are restricted by severe market and production constraints. While these remain, investment in small-scale irrigation will enhance local resilience but is unlikely to be transformative.

The planned expansion of irrigation requires attention to both water resource availability and institutions for water distribution. The focus on resource management and conservation is very positive, and suitable biophysical and institutional approaches will need to be developed at local level.

Water for livestock – in both agricultural and pastoral areas – merits greater attention. Livestock play a vital but little-recognized role in agricultural production, but poor access to water, particularly in the dry season, undermines livestock health and productivity. Better livestock water management would also reduce land degradation. Building on the participatory integrated approach to water and pasture management, seen in some recent investment programmes in pastoral areas, offers considerable promise to enhance resilience in these areas.

Notes

1 Defined by the World Health Organization (WHO) as the sum of years of potential life lost due to premature mortality and the years of productive life lost due to disability.
2 Based on a conservative assumption of 30 minutes time saving per day for improved sanitation facilities and non-household water services, and 90 minutes per day for household piped services.
3 Exchange rate US$1 = ETB 18.02 (17 July 2012) used throughout chapter.

References

Abebe, H. and Deneke, I. (2008) 'The sustainability of water supply schemes: a case study in Alaba Special woreda', *RiPPLE Working Paper 5*, RiPPLE, Addis Ababa.

Abebe, H., Bedru, M., Ashine, A., Hilemariam, G., Haile, B., Dimtse, D., and Adank, M. (2010) 'Equitable water service for multiple uses: a case from Southern Nations, Nationalities and People's Region (SNNPR), Ethiopia', *RiPPLE Working Paper 17*, Research-inspired Policy and Practice Learning in Ethiopia and the Nile Region (RiPPLE), Addis Ababa. All RiPPLE papers available from: <www.rippleethiopia.org/> [accessed July 2012].

Abebe, Y. D. and Geheb, K. (eds.) (2003) *Wetlands of Ethiopia: Proceedings of a Seminar on the Resources and Status of Ethiopia's Wetlands*, IUCN Wetland and Water Resources Programme, Switzerland.

Anderson, E. and Hagos, F. (2008) 'Economic impacts of access to water and sanitation in Ethiopia: evidence from the Welfare Monitoring Surveys', *RiPPLE Working Paper 3*, RiPPLE, Addis Ababa.

Awulachew, S. B., Yilma, A. D., Leulseged, M., Loiskandl, W., Ayana, M., Alamirew, T. (2007) 'Water resources and irrigation development in Ethiopia', *IWMI Working Paper 123*, International Water Management Institute (IWMI), Colombo.

Bach, C. F. and Pinstrup-Andersen, P. (2008) 'Agriculture, growth and employment in Africa – need for institutions, infrastructure, information, innovation, investments and integration', *Discussion Paper Number 2, Civil Society Input to the Africa Commission on Effective Development Cooperation with Africa*. Available from: <www.friisbach.dk/fileadmin/cfb/images/Africa/Agriculture-final.pdf> [accessed July 2012].

Bartram, J. and Cairncross, S. (2010) 'Hygiene, sanitation and water: forgotten foundations of health', *PLoS Medicine* 7(11): 1–9 <http://dx.doi.org/10.1371/journal.pmed.1000367>.

Behnke, R. (2010) *The Contribution of Livestock to the Economies of IGAD Member States: Study Findings, Application of the Methodology in Ethiopia and Recommendations for Further Work*, Livestock Policy Initiative, International Governmental Authority on Development (IGAD), Djibouti.

Calow, R., MacDonald, A., Nicol, A., and Robins, N. (2010) 'Groundwater security and drought in Africa: linking water availability, access and demand', *Groundwater* 48(2): 246–56 <http://dx.doi.org/10.1111/j.1745-6584.2009.00558.x>.

Coulter, L. (2008) *Household Water Economy Analysis in Bale Pastoral Livelihood Zone, Oromiya Region, Ethiopia: Baseline Report*, Livelihoods Integration Unit, Addis Ababa.

Coulter, L., Kebede, S. and Zeleke, B. (2010) 'Water economy baseline report: water and livelihoods in a highland to lowland transect in eastern Ethiopia', *RiPPLE Working Paper 16*, RiPPLE, Addis Ababa.

Curtis, V. and Cairncross, S. (2003) 'Effect of washing hands with soap on diarrhoea risk in the community: a systematic review', *The Lancet Infectious Diseases* 3(5): 275–81 <http://dx.doi.org/10.1016/S1473-3099(03)00606-6>.

Eshetu, S., Belete, B., Goshu, D., Kassa, B., Tamiru, D., Worku, E., Lema, Z., Delelegn, A., Tucker, J., and Abebe, Z. (2010) 'Income diversification through improved irrigation in Ethiopia: impacts, constraints and prospects for poverty reduction', *RiPPLE Working Paper 14*, RiPPLE, Addis Ababa.

Federal Democratic Republic of Ethiopia (FDRE) (2011) *Ethiopia's Climate-resilient Green Economy: Green Economy Strategy*, FDRE, Addis Ababa.

FDRE (2011) *The WASH Implementation Framework (WIF) – Summary*, version: 27 July 2011, FDRE, Addis Ababa.

Gebreselassie, S. (2010) 'Creating new markets via smallholder irrigation: the case of irrigation-led smallholder commercialization in Lume District, Ethiopia', *Working Paper 018*, Future Agricultures Consortium, Brighton.

Giordano, M. (2006) 'Agricultural groundwater use and rural livelihoods in sub-Saharan Africa: a first-cut assessment', *Hydrogeology Journal* 14(3): 310–18 <http://dx.doi.org/10.1007/s10040-005-0479-9>.

Gomes, N. (2006) 'Access to water, pastoral resource management and pastoralists' livelihoods. Lessons learned from water development in selected areas of Eastern Africa (Kenya, Ethiopia, Somalia)', *Livelihood Support Programme Working Paper 26*, Food and Agricultural Organization (FAO), Rome.

Grey, D. and Sadoff, C. (2007) 'Sink or swim? Water security for growth and development', *Water Policy* 9: 545–71 <http://dx.doi.org/10.2166/wp.2007.021>.

Guerrant, R.L., Kosek, M., Lima, A.A.M., Lorntz, B., and Guyatt, H.L. (2002) 'Updating the DALYs for diarrhoeal disease', *Trends in Parasitology* 18(5): 191–3 <http://dx.doi.org/10.1016/S1471-4922(02)02253-5>.

Hagos, F., Boelee, E., Awulachew, S. B., Slaymaker, T., Tucker. J., and Ludi, E. (2008) 'Water supply and sanitation (WSS) and poverty: micro level linkages in Ethiopia', *RiPPLE Working Paper 8*, RiPPLE, Addis Ababa.

Hagos, F., Makombe, G., Namara, R. E., and Awulachew, S. B. (2009) 'Importance of irrigated agriculture to the Ethiopian economy: capturing the direct net benefits of irrigation', *IWMI Research Report 128*, IWMI, Colombo.

Hunter, P. R., MacDonald, A., M., and Carter, R. C. (2010) 'Water supply and health', *PLoS Medicine* 7(11): 1–9 <http://dx.doi.org/10.1371/journal.pmed.1000361>.

Hussain, I. and Hanjra, M. A. (2004) 'Irrigation and poverty alleviation: review of the empirical evidence', *Irrigation and Drainage* 53: 1–15 <http://dx.doi.org/10.1002/ird.114>.

Hutton, G. and Haller, L. (2004) *Evaluation of the Costs and Benefits of Water and Sanitation Improvements at the Global Level*, World Health Organization (WHO), Geneva.

Hutton, G., Haller, L., and Bartram, J. (2007) 'Global cost-benefit analysis of water supply and sanitation interventions', *Journal of Water and Health* 5: 481–502 <http://dx.doi.org/10.2166/wh.2007.009>.

International Fund for Agriculture Development (IFAD) (2009) *Country Programme Evaluation, Federal Democratic Republic of Ethiopia. Report No. 2045-ET*, IFAD, Addis Ababa.

Kemp-Benedict, E., Cook, S., Allen, S. L., Vosti, S., Lemoalle, J., Giordano, M., Ward, J., and Kaczan, D. (2011) 'Connections between poverty, water and agriculture: evidence from 10 river basins. *Water International* 36(1): 125–40 <http://dx.doi.org/10.1080/02508060.2011.541015>.

Ligon, E. and Sadoulet, E. (2007) *Estimating the effects of aggregate agricultural growth on the distribution of expenditures'*, *Background Paper for the World Development Report 2008*, World Bank, Washington, D.C.

Lipton, M. and Litchfield, J. (2003) *Preliminary Review of the Impact of Irrigation on Poverty, with Special Emphasis on Asia*, FAO, Rome.

MacDonald, A. M., Bonsor, H. C., Calow, R. C., Lapworth, D. J., Tucker, J. E., and O'Dochartaigh, B. É. (2011) 'Groundwater resilience to climate change in Africa', *British Geological Survey Open Report*, OR/11/031.

Masiyandima, M. and Giordano, M. F. (2007) 'Sub-Saharan Africa: opportunistic exploitation', In M. Giordano and K. G. Villholth (eds.) *The Agricultural Groundwater Revolution: Opportunities and Threats to Development*, IWMI, Colombo.

McCartney, M. and Smakhtin, V. (2010) 'Water storage in an era of climate change: addressing the challenge of increasing rainfall variability', *IWMI Blue Paper*, IWMI, Colombo.

Ministry of Finance and Economic Development (MoFED) (2010) *Growth and Transformation Plan (GTP) 2010/11–2014/15, Draft of September 2010*, Federal Democratic Republic of Ethiopia (FDRE), Addis Ababa.

Nassef, M. with Belayhun, M. (2012) *Water Development in Ethiopia's Pastoral Areas. A Synthesis of Existing Knowledge and Experience*, Save the Children USA, Addis Ababa, and Overseas Development Institute (ODI), London.

Negash, F. (2012) 'Managing water for inclusive and sustainable growth in Ethiopia: key challenges and priorities', *Background Paper for the European Report on Development, 2011–2012*, European Union, Brussels.

Opio, C. (2009) *Water and Livestock for Rural Livelihoods, InnoWat Topic Sheet*, IFAD, Rome.

Peden, D., Tadesse, G., Misra, A.K., Ahmed, F.A., Astatke, A., Ayalneh, W., Herrero, M., Kiwuwa, G., Kumsa, T., Mati, B., Mpairwe, D., Wassenaar, T., and Yimegnuhal, A. (2007) 'Water and livestock for human development', in D. Molden (ed.), *Water for Food, Water for Life: A Comprehensive Assessment of Water Management in Agriculture,* pp. 485–514, Earthscan, London and IWMI, Colombo.

Perfecto, I. and Vandermeer, J. (2010) 'The agroecological matrix as an alternative to the land-sparing/agriculture intensification model', *PNAS* 107(13): 5786–91 <http://dx.doi.org/10.1073/pnas.0905455107>.

Prüss-Üstün, A., Bos, R., Gore, F. and Bartram, J. (2008) *Safer Water, Better Health: Costs, Benefits, and Sustainability of Interventions to Protect and Promote Health*, WHO, Geneva.

Stockholm International Water Institute (SIWI) (2005) 'Making water a part of economic development: the economic benefits of improved water management and services', *Report for the Commission on Sustainable Development*, SIWI, Stockholm. Available from: <www.who.int/water_sanitation_health/waterandmacroecon.pdf> [accessed 13 July 2012].

Tucker, J. (2009) 'Economic benefits of access to water in Ethiopia: the case for packages of investments', *RiPPLE Briefing Paper 2*, RiPPLE, Addis Ababa.

Tucker, J. and Yirgu, L. (2011) 'Water in food security assessment and drought early warning: experience from sub-Saharan Africa with a special focus on Ethiopia', *RiPPLE Working Paper 21*, RiPPLE, Addis Ababa.

Tucker, J., Calow, R., Coulter, L., and MacDonald, A. (forthcoming) *Exploring Seasonal Drivers of Water Use and their Policy Implications: Quantitative Findings from a Highland to Lowland Transect in Ethiopia Using WELS Data*, London.

United Nations Economic Commission for Africa (UNECA) (2011) *Economic Report on Africa 2011: Governing Development in Africa – the Role of the Stats in Economic Transformation*, UNECA, Addis Ababa.

Van Den Berg, M. and Ruben, R. (2006) 'Small-scale irrigation and income distribution in Ethiopia', *Journal of Development Studies* 42(5): 868–80 <http://dx.doi.org/10.1080/00220380600742142>.

Wilson, R. T. (2007) 'Perceptions, practices, principles and policies in provision of livestock water in Africa', *Agricultural Water Management* 90: 1–12 <http://dx.doi.org/10.1016/j.agwat.2007.03.003>.

World Bank (2006) *Ethiopia: Managing Water Resources to Maximise Sustainable Growth: A World Bank Water Resources Assistance Strategy for Ethiopia*, World Bank, Washington, D.C.

World Bank (2008) *Investment in Agricultural Water for Poverty Reduction and Economic Growth in Sub-Saharan Africa. Synthesis Report* (A collaborative program of African Development Bank – AFDB, FAO, IFAD, IWMI, and the World Bank), World Bank, Washington, D.C.

World Health Organization (WHO) (2009) *Country Profile of Environmental Burden of Disease: Ethiopia*, Public Health and the Environment, WHO, Geneva. <www.who.int/quantifying_ehimpacts/national/countryprofile/ethiopia.pdf> [accessed 13 July 2012].

You, L., Ringler, C., Nelson, G., Wood-Sichra, U., Robertson, R., Wood, S., Guo, Z., Zhu, T., and Sun, Y. (2010) 'What is the irrigation potential for Africa? A combined biophysical and socioeconomic approach', *International Food Policy Research Institute (IFPRI) Discussion Paper 009933*, IFPRI, Washington D.C.

About the authors

Josephine Tucker is a Research Fellow at the Overseas Development Institute (ODI) with a background in ecology and water management and particular experience in Ethiopia and East Africa. Her research interests include water–food–ecosystem linkages, the role of water in resilience and disaster risk management, and equitable governance of water in rural and urban contexts. She is experienced in supporting multi-stakeholder platforms which link research, policy, and practice.

Zelalem Lema is a Research Officer at the International Livestock Research Institute (ILRI) in Ethiopia, where he works on innovation in agricultural systems for enhanced rainwater management. He holds a BSc in Agricultural Extension, and an MSc in Agricultural Economics and Rural Development from Haramaya University; his MSc research focused on willingness to pay for improved rural water supply. Zelalem worked with RiPPLE from 2007–12 in a number of roles: *Woreda* Coordinator for Goro Gutu (where he established and facilitated the local LPA and coordinated action research); Research and Communications Officer; and finally Senior Research and Development Officer, with responsibility for various projects focusing on multiple-use water services, rainwater harvesting, and integrated approaches to urban agriculture and wastewater management.

Samson Eshetu Lemma is a staff member of the Department of Rural Development and Agricultural Extension at Haramaya University and is currently undertaking a PhD at Sokoine University of Agriculture, Tanzania. His research interests include agricultural extension and advisory services, communication, income diversification, and participatory approaches. Samson led the income diversification research under RiPPLE, and has also led the production of a multimedia package module (which is an Open Educational Resource) under the AgShare Project (Michigan State University and Open Educational Resources Africa). He has undertaken a variety of consultancy work and training programmes with a range of NGOs and government organizations.

CHAPTER 7

Responding to climate variability and change: implications for planned adaptation

Lindsey Jones, Lorraine Coulter, Million Getnet Gebreyes, Beneberu Shimelis Feleke, Naomi Oates, Leulseged Yirgu Gebreamlak and Josephine Tucker

Ethiopia faces a difficult task in responding to diverse and variable climate. Like much of sub-Saharan Africa (SSA), the country experiences considerable water stress as a result of complex interactions between various natural, economic, social, and political processes. This complex web makes effective water resource management a considerable challenge; one made all the more difficult in light of changes to the global climate regime. The impacts of climate change are likely to place significant additional pressures on water availability, accessibility, supply, and demand. With this in mind, striking a balance between reducing exposure to the direct impacts of climate change and addressing the root causes of vulnerability (of both climate change and wider development pressures) is a key priority for planned adaptation at various levels of governance – from the local level to the regional and national.

Climate change and the water sector: global and African perspectives

Climate variability and change is a challenge for current and future water management systems. Observed long-term changes in the large-scale hydrological cycle and modelling of climate and water regimes point to considerable impacts on both surface and groundwater systems. At a global scale, observed warming over recent decades is associated with changes in various components of the water system. Modelled simulations of future climates in this century are consistent in projecting regional changes in timing and distribution of precipitation, river run-off, and water availability (Bates et al., 2008).

While broad trends are identifiable, the impacts of climate change will differ from region to region (IPCC, 2007; see also Box 7.1). Projections for Africa indicate that it will experience a sharp warming trend (roughly 2–4.5°C by 2100 for SSA), above the global average (Müller, 2009). With regards to rainfall, much larger uncertainties exist. But models show some agreement that precipitation will decrease over much of northern and southern Africa and increase over West Africa and eastern and central Africa (Goulden et al., 2009). Modelling results also suggest a significant decrease in run-off in northern and southern Africa, with increases in eastern and semi-arid sub-Saharan regions (Arnell, 2004).

> **Box 7.1 Summarized projections of future global climate trends related to freshwater resources**
>
> Synthesized findings from the Intergovernmental Panel on Climate Change's (IPCC's) technical paper on 'Climate Change and Water' include:
> - observed warming over several decades, leading to changes in the large-scale hydrological cycle.
> - an increase in annual average river run-off and water availability at high latitudes and in some wet tropical areas, and a decrease over some dry regions at mid-latitudes and in the dry tropics, by the middle of this century.
> - increased precipitation intensity and variability and an increased risk of flooding and drought in many areas.
> - a decline of water supplies stored in glaciers and snow cover during this century.
> - higher water temperatures and changes in extremes, including floods and droughts, that affect water quality and exacerbate water pollution.
>
> *Source:* adapted and expanded from Bates et al., 2008, in Coulter et al., 2010

Not all impacts will be harmful. In some areas, increased annual run-off may benefit a variety of water users by, for example, increasing renewable water resources available for irrigation (Arnell, 2004). At the global level however, the negative impacts of future climate change on freshwater systems are expected to outweigh the benefits (Bates et al., 2008).

Projections of future climate and its impacts on water resources must be seen in the context of large uncertainties in the modelling process, increasing as we move from modelling impacts of rising temperatures to changing precipitation and from here to impacts on run-off and groundwater. Internal variability and the complexity of the climate system, uncertainty in future greenhouse gas (GHG) and aerosol emissions, uncertainties in relation to land use change or water demand, and difficulties in combining outputs with hydrological models are all challenges in simulating interactions within the global climate and its impacts on water resources.

As a result, there is considerable variation in the outputs of different models. Simulations for some areas across Africa disagree, some even on the direction (positive or negative) of precipitation change (IPCC, 2007). Moreover, projections become less consistent between models at lower spatial scales (Bates et al., 2008). This is particularly problematic as the quality of downscaled model forecasts becomes less reliable at those scales (national or river basin) where decisions on water allocation to different uses have to be made (Ziervogel and Zermoglio, 2009).

Despite these uncertainties, there is enough confidence in both observational records and modelled projections to inform policy-making at higher levels. They provide rich evidence that freshwater resources are vulnerable to climate change, with a whole host of implications for societies and the ecosystems on which they depend that require informed and forward-looking responses.

Ethiopia's climate profile

Ethiopia has a diverse and variable climate. Temperature and rainfall distributions vary across the country, linked to topography, seasonal cycles, and responses to regional and global weather mechanisms (see Figure 7.1).

Ethiopia can be divided into three major topographic regions: the north, central, and south-western highlands and surrounding lowlands; the southeastern highlands and the surrounding lowlands; and the Rift Valley that divides the highlands in two. Ethiopia is divided into five agro-climatic zones, broadly following altitude, rainfall, and temperature with distinct characteristics and specific agricultural outputs and practices (Hurni, 1998):

- *Berha:* below 500 m above sea level (asl), with mean annual temperatures above 25°C and annual rainfall of less than 600 mm;
- *Kolla*: between 500 and 1,500 m asl, with mean annual temperatures between 20 and 28°C and annual rainfall between 600 and 900 mm;
- *Weyna Dega*: between 1,500 and 2,300 m asl, with mean annual temperatures between 16 and 20°C and annual rainfall above 900 mm;
- *Dega*: between 2,300 and 3,200 m als with mean annual temperatures of between 6 and 16°C and annual rainfall above 900 mm;
- *Wurch*: above 3,200 m asl with mean annual temperatures below 6°C and annual rainfall above 1,400 mm.

Most of Ethiopia shares three common seasons, driven largely by the shifting position of the Inter-Tropical Convergence Zone (ITCZ). There is a main rainy season, *Kiremt*, from June to September, when the ITCZ is at its most northern position. Several northern and central parts of Ethiopia also have a short rainy season, *Belg*, from February to May. The southern regions also experience rainfall during the *Belg*, followed by a drier period from October to January called the *Bega*. The eastern corner of Ethiopia receives little rainfall throughout the year (McSweeney et al., 2010a; 2010b).

Strong links exist between the distribution and intensity of rainfall and *El Niño/La Niña*–Southern Oscillation (ENSO) events. In particular, warm *El Niño* phases are thought to reduce rainfall during the *Kiremt* in the northern and central areas of the country, increasing the likelihood of drought, while it also increases rainfall in the *Belg* rainfall season in southern Ethiopia (McSweeney et al., 2010a; 2010b).

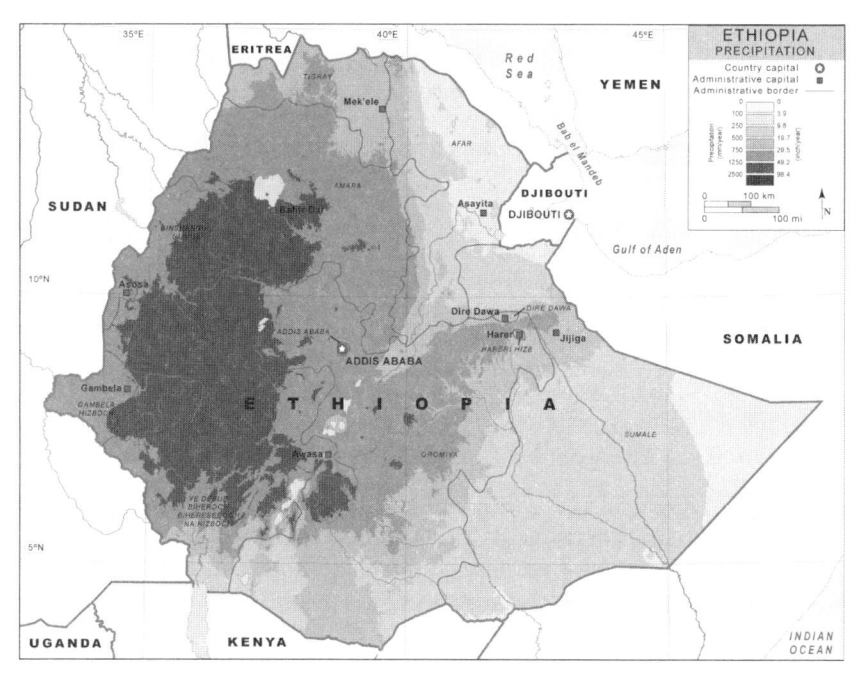

Figure 7.1 Rainfall distribution in Ethiopia

Note: This map does not reflect the recent recognition of South Sudan.

Source: © Copyright 2007 by World Trade Press. All rights reserved.

Ethiopia's changing climate

Strong variations in annual and decadal rainfall patterns and high inter-annual and intra-seasonal rainfall variability over most of Ethiopia make it hard to detect long-term trends, including those due to anthropogenic climate change (Conway and Schipper, 2011). Available information suggests an increase in nationwide annual mean temperatures of 1.3°C between 1960 and 2006 – an average increase of 0.28°C per decade. It also points to the increasing frequency of hot days and nights (McSweeney et al., 2010a; 2010b). Time-series analysis suggests declining March to September rainfall in north-eastern and southern regions, with fairly constant rainfall for north-western Ethiopia (Funk et al., 2005; see also Box 7.2). The overall national trend is more or less constant with no statistically significant changes in rainfall for any season since 1960 (NMA, 2007; McSweeney et al., 2010a; 2010b).

A major challenge in assessing Ethiopia's observational record is a general lack of accurate and reliable data. This problem is exacerbated by large gaps in collated data for many weather stations across the country.

> **Box 7.2 Farmers' perceptions of change**
>
> Participatory research conducted by the Research-inspired Policy and Practice Learning in Ethiopia and the Nile Region (RiPPLE) programme in eastern Ethiopia suggests that farmers' perceptions of temperature trends in the region are generally in line with observed records, identifying distinct shifts in climatic zones, with the *Kolla* (warm zone) replacing the *Weyna Dega* (higher altitude, cooler zone). Farmers also point to an increase in dry spells and temperature, resulting in the disappearance of key fodder species resulting in lower fodder quality and deterioration of livestock (linked also to wider land-use management practices). Farmers also mentioned increased seasonality in rainfall patterns, including changes in timing, intensity, and duration of rainfall.
>
> While participatory research of this sort is subject to various forms of bias, it provides insights into perceptions of seasonal change and can help explain observed coping strategies.
>
> *Source:* Kaur et al., 2010

With regard to projections of Ethiopia's future climate, though limited outputs from models that have more regional relevance are available, decision-makers more often than not have to rely on coarser-resolution climate models such as Global Circulation Models (GCMs).[1] GCMs are complex climate models used to predict future climates using various scenarios to see how the climate will evolve under certain parameters. Aggregate analyses of various GCMs suggest an increase in temperature under various scenarios (see Box 7.3).[2] Projections for rainfall indicate, fairly consistently, an increase in annual rainfall, largely as a result of higher rainfall during *Belg* in southern Ethiopia (McSweeney et al., 2010a; 2010b). Models also suggest more frequent and intense extreme climatic events – both in relation to temperature and precipitation. However, there are variations in the direction or degree of change at the sub-national level and between different seasons (see Box 7.3). Figure 7.2 provides a useful illustration of the coarse nature of GCM model outputs for Ethiopia, and the difficulties in making this information relevant for decision-making at sub-national and local scales.

> **Box 7.3 Summary of GCM ensemble projections for Ethiopia[1]**
>
> *Temperature*
> - The mean annual temperature is projected to increase by 1.1 to 3.1°C by the 2060s, and 1.5 to 5.1°C by the 2090s. Considerable variability exists, as running different models using the same emissions scenario can produce a variety of projections spanning a range of up to 2.1°C.
> - All projections indicate substantial increases in the frequency of days and nights that are considered 'hot' in the current climate.
> - All projections indicate decreases in the frequency of days and nights that are considered 'cold' in the current climate.

Box 7.3 Continued

Precipitation

- Projections from different models in the ensemble are broadly consistent in indicating increases in annual rainfall in Ethiopia.
- Projections of change in the rainy seasons AMJ[2] and JAS[2] that affect the larger portions of Ethiopia are more mixed, but tend towards slight increases in the south-west and decreases in the north-east.
- The models in the ensemble are broadly consistent in indicating increases in the proportion of total rainfall that falls in 'heavy' events. The largest increases are seen in JAS and OND[2] rainfall.
- The models in the ensemble are broadly consistent in indicating increases in the magnitude of 1- and 5-day rainfall maxima.

Notes

1 All projections are relative to a 1970–99 baseline.
2 Seasons are described as sequences of months (i.e. AMJ represents April, May, June, JAS represents July, August, September, and OND represents October, November, December).

Source: McSweeney et al., 2010a, 2010b

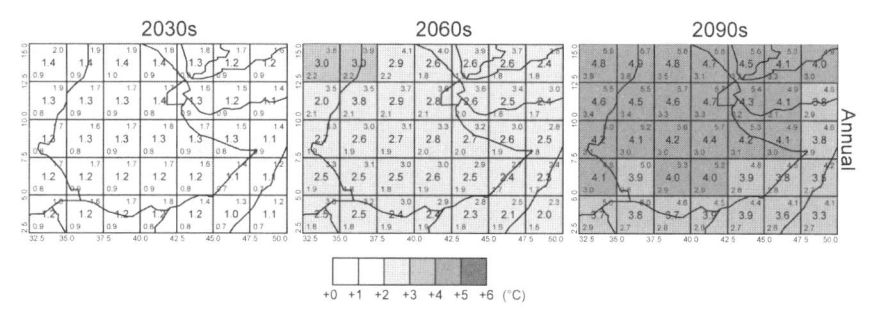

Figure 7.2 Coarse-resolution GCM outputs showing projected changes in temperature at a sub-national scale

Note: Spatial patterns of projected change in mean annual and seasonal temperature for 10-year periods in the future under the A2 emissions scenario. All values are anomalies relative to the mean climate of 1970–99. In each grid box, the central value gives the ensemble median and the values in the upper and lower corners give the ensemble maximum and minimum.

Source: McSweeney et al., 2010a; 2010b

Exploring vulnerability to climate change: adding a layer of complexity to existing development challenges

Knowing how the climate may change is important, but it is only part of the picture. To understand how differences in temperature and rainfall will impact on ecosystems and the livelihoods they support, their impacts need to be examined in the context of existing development pressures, such as population growth, changing land use practices, and natural resource depletion (Urama and Ozor, 2010; Magrath, 2008; Nori and Davies, 2006; Jones et al., 2010a). With regard to impacts on the water sector, drivers of demand, including population growth, land use change, economic growth, and technological change will all have an impact on the water cycle. For example, Ethiopia's population is expected to increase from more than 80 million in 2010 to around 146 million by 2050. The pressure this generates on land and water is likely to dwarf the impacts of climate change in the near-to-medium-term future (Calow and MacDonald, 2009). In many contexts, climate change factors can be considered as 'merely items to be included in a risk assessment alongside other risks to the sustainability of water and sanitation services' (WSP, 2010: 11). Nevertheless, climate change is likely to have significant effects on efforts to address and achieve water security, in SSA and Ethiopia in particular.

To date, issues of vulnerability, livelihood, and water security are particularly pronounced in Ethiopia as rainfed agriculture has formed the backbone of the economy, with irrigation applied only to five per cent of irrigable land (World Bank, 2006). This dependency on rainfed agriculture means that Ethiopia's economy is extremely sensitive to the adverse impacts of varied and extreme weather (Giorgis et al., 2006), which is exacerbated by lacking market mechanisms which could buffer the economic impacts of climate variability on agricultural production (Grey and Sadoff, 2007).

Dependency on rainfed agriculture, limited transport and storage infrastructure, and a lack of effective risk management mechanisms at household and national levels have resulted in a close correlation between Ethiopia's economic growth pattern and rainfall deviations. The economic cost of hydrological variability has been estimated at over one-third of the nation's average annual growth potential (World Bank, 2006). Conway and Schipper (2010) point out, however, that the historical correlation between gross domestic product (GDP) and rainfall becomes weaker after 2000, due primarily to greater diversification of Ethiopia's economy, and the fact that recent droughts have been concentrated mainly in southern and south-eastern Ethiopia, areas with limited contribution to the GDP (Ludi et al., 2011).

The dependency of Ethiopia's economy on rainfall is exacerbated by a lack of water storage infrastructure and institutions to mitigate hydrological diversity (Grey and Sadoff, 2007). Though Ethiopia has relatively abundant water resources, it has one of the lowest reservoir storage capacities in the world, at just 50 m³ per person, compared with 4,700 m³ in Australia (UNDP, 2007). Given projections of increased variability of rainfall and reduced availability of water resources, as well as food security issues, Ethiopia requires significant investments in water resource and storage development, particularly for irrigation (Calow et al., 2002).

Domestic water services, many of which depend on shallow groundwater, are also affected by rainfall variability. Sustainability problems are likely to increase vulnerability for much of Ethiopia's rural population, with many households already today receiving poor or non-existent services. Much of this is due either to restricted seasonal supply or malfunctioning systems (see Chapter 5).

Vulnerability to water stress and poor access to domestic and productive water have been demonstrated in a RiPPLE study in eastern Ethiopia (Coulter et al., 2010; Tucker et al., forthcoming). Research was conducted along a highland to lowland transect in Oromia Region, using a modified Household Economy Approach (HEA; see FEG et al., 2008) adapted to include water, termed Water Economy for Livelihoods (WELS). Besides assessing how people access food and income, and what puts different households at risk of food shortages, WELS adds further insights on water use volumes and the labour demands of water collection.

Three livelihood zones were selected:

- Highland 'Wheat Barley Potato' (WBP) zone, characterized by crop-based livelihoods and high population density, with unprotected springs as the dominant water source.
- Midland 'Sorghum Maize Chat' (SMC) zone, an agricultural zone with lower population density, more livestock, and water provided by a combination of scattered springs, seasonal ponds, and a few permanent rivers.
- Lowland 'Shinile Agropastoral' (SAP) zone, where livestock provide most cash income. Humans and livestock mostly rely on deep groundwater that can only be accessed via boreholes, a few permanent rivers, and seasonal pools and ponds in the wet season.

Results show that levels of water use are below international minimum standards across the three zones. Volumes used are particularly low for poorer households, and in the dry season (Figure 7.3). Dry season water use is restricted by a combination of more difficult access (some sources dry up or fail, leading to greater distances travelled for water and longer queues at remaining sources) and labour constraints, particularly for poor households that tend to be smaller and rely heavily on wage labour (for which demand often peaks during the dry season). Poor households are also unable to buy water pumps, unlike better-off households.

For some households the time spent on collecting water comes at a high price in terms of lost livelihood opportunities, with trade-offs having to be

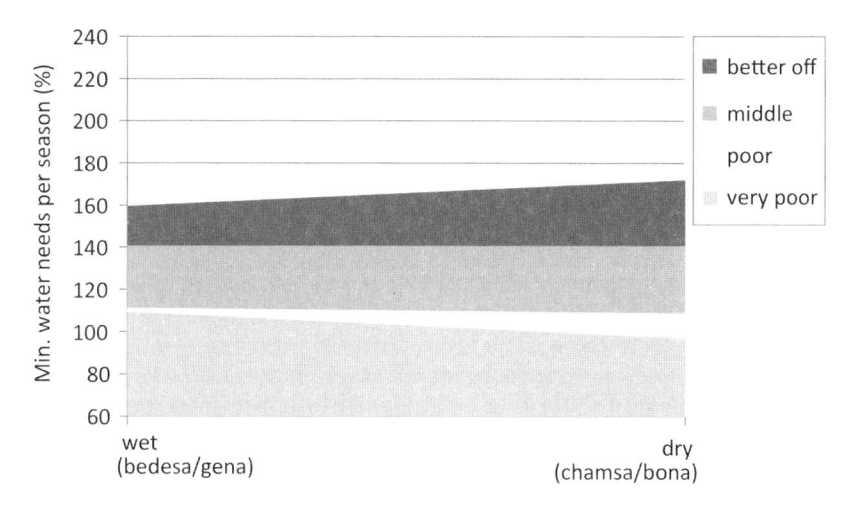

Figure 7.3 Water use for human consumption as a percentage of minimum human needs in the highland Wheat Barley Potato (WBP) livelihood zone

Source: Coulter et al., 2010

made between water collection and income-generating or agricultural activities. Poor water access impacts water for livestock particularly badly: many households are unable to provide enough water for livestock in the dry season (Figure 7.4). Poor households in the agropastoral lowland zone report that this affects livestock condition and that they receive prices that are 20 per cent lower for their cattle than better-off households. Although this difference may not be due to water shortages alone, inequities in water access contribute to vicious cycles of wealth inequality. In midland and highland zones, where poor households have few livestock, it is the relatively better-off households who struggle to provide water for their larger herds. This may increase pressure further on existing water infrastructure.

With water use at such low levels even in a typical 'non-drought' year, poor households are highly vulnerable to any further deterioration in water access. Without investment to provide more improved and reliable sources, many poor households will be forced to reduce water use further (affecting human and livestock health) or sacrifice wage-earning or food production opportunities to collect water. Relatively better-off households tend to be more dependent on water-based livelihood strategies such as livestock and sometimes irrigated crop production and will, therefore, also be vulnerable to climate change hazards that increase water stress.

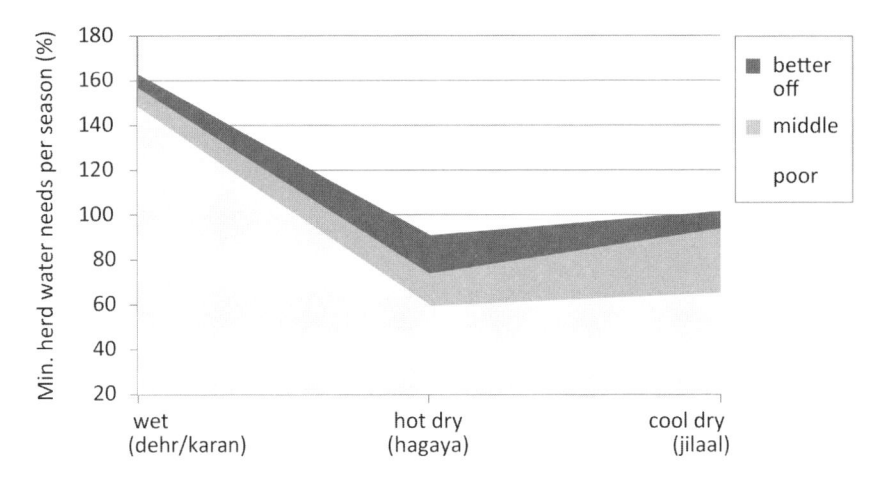

Figure 7.4 Water use by livestock in the lowland Shinile Agropastoral (SAP) livelihood zone as a percentage of minimum herd needs

Source: Coulter et al., 2010

Responding to climate variability and change

Actions to respond to climate change are extremely varied. Some will be planned, often with outside support by local and national governments, NGOs, civil society or private sector entities. Most will, however, occur at the household and community levels without any direct assistance from external formal institutions. They are rarely a conscious response to a changing climate itself but rather a reaction to the economic or socio-political consequences of the climate condition. Examples of such autonomous and short-term actions – or coping strategies – include: temporary pastoral migration and sale of livestock; regulation of water resources through local water committees and institutions; and changes in agricultural practices or crop varieties.

These actions are nothing new; households in rural Ethiopia have, over centuries, developed various strategies to cope with climatic shocks or stresses and have adapted their livelihoods to changing environmental conditions. The ability of a household or community to cope with and adapt to climate shocks and stresses depends on a range of physical, economic, and social determinants (Brooks et al., 2005; Jones et al., 2010a). Some strategies, however, can undermine future ability to adapt, as they erode household or community assets. The term 'adaptive capacity' is helpful in understanding the complex process involved in responding to climate events, and refers to the ability of a system (often a community or household) to anticipate, deal with, and respond to change – whether it is climate change or wider development pressures (Levine et al., 2011). Importantly, the term refers to the *potential* to adapt, and not necessarily the act of adapting as an output.

In Ethiopia, the capacity of rural communities to both cope with and adapt to a changing external environment depends on factors such as local natural resources, access to livelihood assets, kinship networks, access to information, skills, and local institutions, amongst others (Ludi et al., 2011). At a national level, Ethiopia's ability to adapt to climate change is limited by limited financial reserves; a lack of technical expertise and information; low access to education, training, health facilities, financial resources, and services; and limited availability of infrastructure and markets among others.

Although most rural Ethiopian communities suffer from low levels of adaptive capacity (Amsalu and Adem, 2009) there are notable differences both within and between them. For example, relatively rich households with a diverse asset base, a well-developed social network, and significant political power tend to be better able to respond and adapt than poorer and more marginalized households (Kaur et al., 2010).

Disparities in levels of adaptive capacity exist not only in relation to assets, but to entitlements, social exclusion, and institutions (Jones and Boyd, 2011). For example, marginalization and inequality can be significant barriers to adaptation for women, children, and various excluded groups across Ethiopia (Ludi et al., 2011).

Not every action taken to adapt to climate variability and change will be successful (Eriksen et al., 2011). Indeed, many adaptation actions focus on short-term benefits and can increase vulnerability in the longer-term (Berrang-Ford et al., 2011). In the Shinile and eastern Hararghe areas, coping strategies adopted by poor households, which run the risk of eroding household and community assets, include the sale of assets and the killing of calves. Household members are also forced to travel long distances to collect water from reliable sources and often reduce their water use for sanitation (Kaur et al., 2010). Other coping strategies frequently observed include cultivating or grazing unsuitable areas, making charcoal, and selling fuel-wood.

Overgrazing and deforestation lead to land degradation and a depleted natural resources base (Amsalu and Adem, 2009). If households have to fall back more frequently on short-term coping strategies because of increased climatic stress, household and communal assets (natural, physical or human) may eventually erode with negative implications for household health, income and food production, and children's education. All of these may then undermine current livelihoods and future ability to adapt to climatic shocks, stresses or change.

This is particularly relevant given that current responses to climate hazards are informed largely by historic practices and past climate experience. Interventions to support successful and sustainable adaptation thus need to be based on longer-term planning and information, and be able to prevent households falling back on short-term coping strategies so as to prevent further asset depletion even in the face of increasing extreme events (Jones et al., 2010a).

Planned adaptation

Planned adaptation at the national level

In Ethiopia, the responsibility for coordinating climate change responses at the national level lies primarily with the Environmental Protection Authority (EPA). As a signatory of the United Nations Framework Convention on Climate Change (UNFCCC), Ethiopia has developed a National Adaptation Programme of Action (NAPA) (summarized in Box 7.4).

Box 7.4 National Adaptation Programme of Action (NAPA)

Ethiopia's NAPA identifies the support required to address urgent and immediate needs and concerns related to adaptation. Select prioritized projects relevant to the water sector in the NAPA include:

1. promoting the drought/crop insurance programme;
2. strengthening/enhancing drought and flood early warning systems;
3. development of small-scale irrigation and water harvesting schemes in arid, semi-arid, and dry sub-humid areas;
4. improving/enhancing rangeland resource management practices in the pastoral areas;
5. community-based sustainable utilization and management of wetlands in selected areas.

Source: UNFCCC

The NAPA's objective was to identify the key regions, sectors, and livelihoods in Ethiopia that are the most vulnerable to climate change and to determine priorities for immediate action (NMA, 2007). With an emphasis on community participation, the projects prioritized for funding focus broadly on natural resources management, irrigated agriculture and water harvesting, and disaster prevention (early warning systems and awareness raising), in addition to human and institutional capacity building.

Ethiopia's NAPA is far from comprehensive in its treatment of climate risks and while it is a first attempt to integrate adaptation and development, it was never intended as a long-term strategy. For example, although several of the proposed measures directly relate to the water sector, certain sub-sectors that matter for community-level adaptation, as they directly contribute to alleviating household poverty and vulnerability, such as domestic water supply and sanitation (WASH), are not addressed (Oates et al., 2011).

More recently, the government has developed the Climate-Resilient Green Economy (CRGE) strategy. The objective is to mainstream adaptation and mitigation across key sectors over the medium to long term, addressing climate change within routine development planning (Oates et al., 2011).

Identifying over 60 initiatives across various sectors, the green economy strategy rests on four pillars:

- 'Improving crop and livestock production practices for higher food security and farmer income while reducing emissions;
- Protecting and re-establishing forests for their economic and ecosystem services, including as carbon stocks;
- Expanding electricity generation from renewable sources of energy for domestic and regional markets;
- "Leapfrogging" to modern and energy-efficient technologies in transport, industrial sectors, and buildings' (FDRE, 2011: 2).

Four initiatives, which could have significant implications for water resources, have been identified for fast-track implementation because of their assumed contributions to growth and their estimated abatement potential, but also because they are expected to attract climate finance for their implementation: exploiting Ethiopia's hydropower potential, large-scale promotion of advanced rural cooking technologies, efficiency improvements in the livestock sector, and reducing emissions from deforestation and forest degradation (REDD).

Meanwhile, the government is also developing a climate resilience strategy to reduce Ethiopia's vulnerability to extreme climate events such as droughts and floods. The extent to which these activities will support adaptation across relevant sectors will depend in part on the ability to develop institutional capacity to implement the strategy, in particular to coordinate planning and implementation of inter-sectoral activities across different ministries, as well as to mobilize sufficient levels of financial support for its delivery.

As the NAPA and CRGE initiatives demonstrate, there is strong interest in climate change issues at high levels of the Ethiopian Government. The Prime Minister plays an active role in international negotiations and in shaping the national agenda (Oates et al., 2011). However, despite clear signs of engagement and strategic-level planning, there appears to have been little progress in mainstreaming adaptation into development practice (Oates et al., 2011; Schipper, 2007b). For example, Conway and Schipper (2011) find that although historic climate trends are considered in project design, the impacts of future climate variability and change are not adequately addressed, and that underlying factors that shape local vulnerabilities are not always recognized. Integrating adaptation considerations into routine development planning would be particularly beneficial for Ethiopia's water sector, for several reasons. First, there is a need to tackle the underlying socio-economic causes of vulnerability to water-related climate hazards and to 'climate-proof' developments to cope with current climate variability, regardless of future change (Schipper, 2007b). Second, water is the primary medium through which climatic changes will be experienced and adaptive capacity could be increased by investing in the development of infrastructure and institutions for water management (Hedger and Cacouris, 2008; IUCN et al., 2009). Third, 'the systemic nature of water' (IUCN et al., 2009: 12) means that decision-makers need to think (and act) beyond traditional sectoral boundaries. Fourth, for a large part, activities in the water sector are inter-sectoral and thus require joint planning among Ministries of Water and Energy, Agriculture, Forestry, Environment, and others.

Despite ample opportunities for development and adaptation 'win–wins', the process of integration is not as simple as it seems. Factors that are well-known in development practice will also be relevant for adaptation activities (Oates et al.,2011):

- inadequate information systems and patchy availability of hydrological and meteorological data;
- underdeveloped communication and coordination mechanisms across sectors, with many aspects of water management outside the influence or control of the Ministry of Water and Energy (MoWE);
- a relatively poor understanding and awareness of climate change issues at lower administrative levels, and a tendency for 'top-down' adaptation planning;
- weak institutions for water management, and a particular lack of technical and financial capacity to address climate change.

Planned adaptation at the local level

Given the significant challenges to national adaptation planning, the value of locally-targeted initiatives must not be downplayed. RiPPLE case studies on options for planned adaptation highlight the fact that the most appropriate solutions are those tailored to a particular local context. Adaptation planning needs to be flexible and attuned to local needs and capabilities, recognizing that communities and individuals can be highly innovative, given the right enabling environment (a fact often overlooked by development interventions) (Jones et al., 2010a).

Examples of local water-based interventions and their effectiveness in reducing vulnerability to climate variability and change have emerged from a RiPPLE case study (Kaur et al., 2010), which carried out impact and adaptation assessments on four interventions identified in the NAPA. The assessments focused on the evaluation of interventions in terms of their effectiveness in supporting adaptation through water resource management and strengthening adaptive capacity at the household level:

- **Small-scale irrigation:** small-scale irrigation based on surface water is a supply side intervention that is highly location specific. Findings suggest that small-scale irrigation can create assets for some wealth groups and enhance the coping capacity of beneficiaries. However, the research suggests that investments have not taken into account issues of changing rainfall variability, extreme rainfall, and long-term climate change and its potential impacts on run-off. Small-scale irrigation schemes based on groundwater sources – where available and accessible – would create

and enhance the community asset base, building local capacity to cope with climate impacts and reduce exposure to climate change. However, issues of equitable access must be taken into account when promoting this option.
- **Rangeland management:** introduction of drought-resistant fodder species and effective management of invasive species have enhanced the productivity of rangelands and contributed to protecting livelihood assets during times of climate stress. Supporting local management practices has played a role in local conflict management mechanisms, though successful resolution is often dependent on recognizing complex social and cultural institutional arrangements.
- **Productive Safety Net Programme (PSNP):** within the study location, the PSNP has created, protected, and enhanced the asset base of local communities in the face of climate stress, but findings also suggest that targeting of both beneficiary areas and individuals should be undertaken carefully to reach the most vulnerable. Most importantly, the PSNP itself needs to be climate-proofed, ensuring that its work programmes are sustainable and effective given projected changes in future climate, and the need for an informed long-term approach to governance and decision-making processes.
- **Multiple-use services (MUS):** examples of MUS interventions in the study sites include investments in water services and infrastructure, incorporating both domestic and productive uses of water, and training in water and sanitation management. Findings suggest that MUS programmes can support the creation and enhancement of assets, and enhance existing coping capacities during times of climate stress. Health and sanitation training provided under the initiative has helped to reduce the incidence and spread of water-borne disease. MUS investments (such as irrigation schemes) should, where effective and sustainable, explore the development of groundwater resources to reduce exposure to climate stress and secure more reliable water resources. Watershed protection needs to be implemented alongside any investment in MUS, making strong sectoral coordination at local level in planning, design, and implementation, a requirement

The study further concluded that:

1. local communities perceive a change in climate in terms of increases in temperature; changes in humidity; soil moisture content and wind direction; and, above all, changes in precipitation patterns;
2. climate-induced changes are likely to impact on the economic and domestic use of water. The impacts are likely to be greater as a result of extreme weather events than climate variability, as there is greater capacity to cope with variability;

3   local coping capacity to deal with climate variability may be undermined by the increased uncertainty posed by climate change, and is insufficient to cope with extreme weather events;
4   the assessed adaptation interventions build on, and fill gaps in, coping capacity – besides considering financial feasibility, they need to be climate-proofed and need to take aspects of equity into consideration.

Though the study focused on a small area of Ethiopia, it sheds valuable light on the complexity of identifying and evaluating the success of different adaptation strategies. No one option is most suitable across Ethiopia's many and diverse contexts. What is urgently needed is effective and practical processes and the required information bases to support the identification of the most appropriate adaptation alternatives tailored to particular local contexts.

Supporting adaptation at different levels of governance: options and implications for policy

Planned adaptation can happen at various levels: from the individual and household level up to the regional, national, and international. Each has its own role in promoting effective and sustainable adaptation; each also has its own barriers and limitations to delivery.

In addition to dealing with current impacts of climatic variability on the hydrological system and its impacts on existing water infrastructure, service delivery, and consumption patterns, resource managers will have to play an active role in adapting design and operational activities to be able to deal with future impacts of climate change. Importantly, though, while the availability of relevant climate information and projections can be of great use to inform adaptation, climate information *per se* is not required to successfully adapt. More important is that other components of adaptive capacity, such as adequate governance systems supporting forward-looking and flexible planning, are put in place.

Changing precipitation patterns, increasing climate variability, and shifting levels of evapotranspiration all suggest that water management practices, which in many cases are not robust enough to deal with current conditions, will be even less able to cope with the impact of climate change. Its effects are likely to have implications for water supply reliability, flood risk, health, agriculture, energy, and aquatic ecosystems (Bates et al., 2008: 16). These impacts cut across sectors, and sectoral responses need, therefore, to be integrated and holistic to enable climate-proofing (OECD, 2009).

Ethiopian and African policy-makers tasked with responding to climate variability and change must act to support adaptation across all scales, despite enormous uncertainties. In addition to the large unknowns in the science of climate change, most African governments are limited in their capacity to deliver effective adaptation action, given their over-stretched financial, institutional, and technical capacities and resources (Urama and Ozor, 2010). Even so, various policy responses can facilitate successful and sustainable adaptation in relation to water and livelihood security, from the local to the national scale. Drawing on RiPPLE's experiences and research, seven strategies are identified to enhance the capacity of the water sector to respond to changing climate and development pressures. These strategies are tailored to an Ethiopian context, but resonate equally across much of Africa:

Investing in information systems

Paucity of reliable information and data for planning, design, and evaluation of policy and practice in the water sector is a major problem. Without a sound information base, including information on hydrogeology, spatio-temporal water availability and use, meteorology, investments in water services and infrastructure, both for domestic or productive uses, run a high risk of failure. We know that the Ethiopian economy is highly vulnerable to climatic variability and extremes, not least because of its high dependence on rainfed agriculture, but also because of the very limited storage, conveyance, and distribution infrastructure. This is not only true with regard to future climate change scenarios, but presents a major obstacle to sound and sustainable development even under current conditions. Without major investments in more reliable information systems, planning for future uncertainty is likely to be constrained and may run the high risk of leading to wasted resources and maladaptation.

Sustainable development of groundwater for enhanced resilience

There is considerable potential for groundwater development to buffer climate variability and change, helping to balance the negative effects of climate change on surface water variability (Döll, 2009). However, future groundwater development, particularly when accessed for productive uses such as irrigated agriculture, will only be sound if evidence-based and cognisant of different possible climate futures. Studies suggest that even if rainfall were to decrease dramatically, recharge will still occur in most years in areas where annual rainfall currently exceeds 500 mm, and the modest recharge requirements of handpump-based domestic supply would be easily satisfied (Calow and MacDonald, 2009; Calow et al., 2002).

While groundwater can play a useful role in enhancing resilience during times of climate stress, questions remain over the sustainability and thresholds for exploitation and management as well as technical and institutional capacities among stakeholders at multiple scales to access groundwater as a sustainable adaptation solution (Coulter et al., 2010).

The current focus on developing groundwater for domestic and productive purposes, often justified on the grounds of enhancing resilience, will only be successful if based on a sound understanding of hydrogeological conditions, and if accompanied by well-designed investments in watershed protection and overall resource management. Both are currently lacking, enhancing the chance of investments being unsustainable in the best case, a maladaptation and waste of resources in the worst case.

Need for diversified storage options

Given the likely increase in rainfall variability and seasonal uncertainty as a result of climate change, water storage provides a useful mechanism for dealing with variability which, if based on sound evidence and planned and managed correctly, increases water security, agricultural productivity, and adaptive capacity (McCartney and Smakhtin, 2010). It is crucial that storage options reflect the diversity and flexibility needed to deal with evolving pressures, and to build redundancy in light of failures within service delivery infrastructure. Findings from RiPPLE's research support the notion that decentralized and varied storage options, at a range of scales, are needed, with better use of natural storage at the point of demand (MacDonald et al., 2008).

'Climate-proofing' the agricultural sector and food system

Insights from the WELS analysis point to the need to support 'climate-proofing' agriculture. However, regardless of how climate change manifests itself in the future, non-climatic factors associated with agricultural productivity (lacking inputs and improved seeds and breeds, declining soil fertility, weak markets, ineffective extension systems, etc.) can be far more damaging than climatic factors in their implications for household livelihoods, particularly over one or two decades. There is still scope to improve yields in countries such as Ethiopia where production is 10 per cent or less of its theoretical potential. 'At the farm level', write Brown and Funk (2008: 611), 'land modifications, in-soil vegetative material and well-placed biodiversity can all play a role in countering unfavourable climatic events' – and in addressing underlying factors of vulnerability. Increased investments in agricultural research and extension, in resource conservation, and in infrastructure and institutions (including those related to water resources management and water service provision) would all help to improve agricultural prospects for Ethiopia's rural populations and could contribute to increased food security at a national level as well.

Given Ethiopia's variable climate, an early warning and early response system that is capable of dealing with extremes such as droughts and floods is required. As part of this, more attention should be paid to learning from past experience, emphasizing the importance of water in drought preparedness and response, and establishing stronger links between those working on water service delivery and those involved in emergency preparedness and response (Tucker and Yirgu, 2010).

Investing in irrigation

Evidence from RiPPLE research suggests that the contribution of small-scale irrigation to income diversification and livelihood resilience can be significant (Eshetu et al., 2010) (see Chapter 6). Given Ethiopia's high dependence on rainfed agriculture, low levels of water exploitation (less than five per cent of total renewable water resources are withdrawn annually), and poor irrigation coverage (roughly five per cent of Ethiopia's land is irrigated), there remains considerable scope to expand irrigation infrastructure and coverage. This is on the proviso that technical skills and the required evidence base for sound planning, design, and implementation are available, including the knowledge of which water sources will be particularly vulnerable to impacts of climate variability and change, as well as local capacities to regulate and manage schemes. In this role, PSNP works could be targeted to fill critical infrastructure gaps such as access to irrigation, road links with market centres or watershed protection measures.

Any such support must ensure that irrigation development is accompanied by complementary investments in market development, transport, and communication infrastructure, and above all sound management of natural resources. Without these, irrigation development is likely to yield limited returns and may not generate the desired rural growth and enhancement of adaptive capacity. Particular emphasis is needed on ensuring that:

- water sources themselves are resilient to a variable climate;
- water investments and systems are proofed against extreme events;
- regulation, management, and maintenance of systems are assured and mechanisms for equitable water distribution and benefit sharing are put in place.

Role of wider development interventions in supporting adaptive capacity

Findings from RiPPLE's research support the notion that climate change cannot be separated from wider development processes. The impacts of climate change on the water sector are only one of many pressures influencing the hydrological cycle and water resource management. Development challenges such as land use change, natural resource degradation, and population dynamics will have significant impacts on water resources and water service delivery and are, in turn, likely to be impacted by the effects of a changing climate (Nicol and Kaur, 2009).

Similarly, research suggests that the ability to adapt to change – whether climate related or otherwise – is linked closely to underlying causes of vulnerability, such as poverty and inequality, rapid population growth, a degraded resource base, and weak systems of governance (Ludi et al., 2011). Many existing development interventions, such as disaster risk reduction (DRR), social protection, and livelihoods approaches, will, therefore, be critical to

Holistic and integrated policy responses

Taking into consideration the wide ranging and cross-sectoral impacts of climate change, adaptation policy has to ensure a holistic approach, recognizing the diversity of overlapping challenges affecting the water sector. An important component of a sound adaptation policy is to ensure an adequate balance of 'hard' and 'soft' adaptation actions, dealing not only with the impacts of climate variability and change but equally with the drivers of vulnerability. The adoption of forward-looking governance and decision-making processes is essential, especially in contexts where climate change will necessitate transformation of livelihoods as opposed to only incremental adaptation (Pelling, 2010). Such an inclusive and anticipatory process is imperative to prevent maladaptation and ensure the sustainability of current and future practices (Jones et al., 2010a).

Given the cross-sectoral nature of climate change impacts, integration of adaptation and sustainable development agendas within government must involve all associated sectors and ministries as well as active and effective participation, coordination, and communication across private, public, and civil society actors. Adaptation has, to date, tended to be characterized as an environmental issue (Schipper, 2007a). Reflecting this, responsibility for climate change in Ethiopia has been devolved recently to the Environmental Protection Authority. Yet, reducing climate change to an 'environment issue' is problematic if the broader development implications of climate risk are not fully recognized. This may lead to adaptation efforts becoming 'additional components' and separate from ongoing development activities (Oates et al., 2011).

Conclusion

Insights from RiPPLE, and wider experiences, show that planned adaptation is already taking place at national and local scales, with significant effects on water resources and the livelihoods that depend on them. The role of policy-makers is to ensure that existing and future actions to facilitate adaptation are inclusive, evidence-based and well coordinated. Despite existing knowledge gaps, it is essential that decisions made in view of adaptation are forward-looking to reduce the risk of maladaptation while remaining flexible enough to allow adjustments as they may become necessary.

These seven options offer solutions in working towards effective delivery of adaptation in Ethiopia's water sector. Though many appear self-evident and not very different from normal development activities in the sector, their delivery will be far from easy. Each will require a mixture of incremental and transformative changes in reconsidering how adaptation is planned, operationalized, delivered, and integrated into on-going development activities.

Ensuring that the central principles are reflected in policy and programming in an informed and forward-looking manner, and that efforts are made to maintain momentum, will be crucial to ensure that policy processes have the greatest impact on adaptation at the local level.

Notes

1 Outputs from various regional climate models are available, but the extent of large uncertainties means that their use in guiding pro-active adaptation at the local level is questionable.
2 Results are shown using an ensemble of 15 GCM models. Scenarios are based on the full range of IPCC emissions scenarios (A2, A1B, and B1).

References

Amsalu, A. and Adem, A. (2009) 'Assessment of climate change-induced hazards, impacts and responses in the southern lowlands of Ethiopia', *FSS Research Report No.4*, Forum for Social Studies (FSS), Addis Ababa.

Arnell, N.W. (2004) 'Climate change and global water resources: SRES emissions and socio-economic scenarios', *Global Environmental Change*, 14(1): 31–52 <http://dx.doi.org/10.1016/j.gloenvcha.2003.10.006>.

Awulachew, S. B., Merrey, D. J., Kamara, A. B., Van Koppen, B., Penning de Vries, F., and Boelee, E. (2005) 'Experiences and opportunities for promoting small-scale/micro irrigation and rainwater harvesting for food security in Ethiopia', *IWMI Working Paper 98*, International Water Management Institute (IWMI), Addis Ababa.

Bates, B. C., Kundzewicz, Z.W., Wu, S., and Palutikof, J.P. (eds.) (2008) *Climate Change and Water: Technical Paper of the Intergovernmental Panel on Climate Change Intergovernmental Panel on Climate Change (IPCC)*, IPCC Secretariat, Geneva.

Berrang-Ford, L., Ford, J.D., and Paterson, J. (2011) 'Are we adapting to climate change?' *Global Environmental Change*, 21(1): 25–33 <http://dx.doi.org/10.1016/j.gloenvcha.2010.09.012>.

Brooks, N., Adger, N., and Kelly, M. (2005) 'The determinants of vulnerability and adaptive capacity at the national level and the implications for adaptation', *Global Environmental Change Part A*, 15(2): 151–63 <http://dx.doi.org/10.1016/j.gloenvcha.2004.12.006>.

Brown, M.E. and Funk, C.C. (2008) 'Food security under climate change', *Science*, 319(5863): 580–1 <http://dx.doi.org/10.1126/science.1154102>.

Calow, R. C. and MacDonald, A. M. (2009) 'What will climate change mean for groundwater supply in Africa?' *ODI Background Note*, Overseas Development Institute (ODI), London.

Calow, R., MacDonald, A., Nicol, A., Robins, N., and Kebede, S. (2002) 'The struggle for water: drought, water security and livelihoods', *British Geological Survey Commissioned Report CR/02/226N*, British Geological Survey, Nottingham.

Conway, D. and Schipper, E.L.F. (2011) 'Adaptation to climate change in Africa: challenges and opportunities identified from Ethiopia', *Global*

Environmental Change 21(1): 227–37 <http://dx.doi.org/10.1016/j.gloenvcha.2010.07.013>.

Coulter, L., Kebede, S., and Zeleke, B. (2010) 'Water economy baseline report: water and livelihoods in a highland to lowland transect in eastern Ethiopia', *RiPPLE Working Paper 16*, RiPPLE, Addis Ababa. All RiPPLE papers available from: <www.rippleethiopia.org/> [accessed July 2012].

Döll, P. (2009) 'Vulnerability to the impact of climate change on renewable groundwater resources: a global-scale assessment', *Environmental Research Letters*, 4(3): 035006 <http://dx.doi.org/10.1088/1748-9326/4/3/035006>.

Eriksen, S., Aldunce, P., Bahinipati, C.S., Martins, R.D., Molefe, J.I., Nhemachena, C., O'Brien, K., Olorunfemi, F., Park, J., Sygna, L., and Ulsrud, K. (2011) 'When not every response to climate change is a good one: identifying principles for sustainable adaptation', *Climate and Development*, 3(1): 7–20 <http://dx.doi.org/10.3763/cdev.2010.0060>.

Eshetu, S., Goshu, D., Kassa, B., Worku, E., Delelegn, A., Tucker, J., Belete, B., Tamiru, D., Lema, Z., and Abebe, Z. (2010) 'Income diversification through improved irrigation in Ethiopia: impacts, constraints and prospects for poverty reduction: evidence from East Hararghe Zone, Oromia Region, Ethiopia', *RiPPLE Working Paper 14*, RiPPLE, Addis Ababa.

Federal Democratic Republic of Ethiopia (FDRE) (2011) *Ethiopia's Climate-resilient Green Economy: Green Economy Strategy*, FDRE, Addis Ababa.

Food Economy Group (FEG) Consulting, Save the Children UK (SCUK), and Regional Hunger and Vulnerability Programme (RHVP) (2008) *The Practitioner's Guide to the Household Economy Approach*, FEG, SCUK, and RHVP. Available from: <www.feg-consulting.com> [accessed July 2012].

Funk, C., Senay, G., Asfaw, A., Verdin, J., Rowland, J., Korecha, D., Eilerts, G., Michaelsen, J., Amer, S., and Choularton, R. (2005) 'Recent drought tendencies in Ethiopia and equatorial subtropical Eastern Africa', *Vulnerability to Food Insecurity: Factor Identification and Characterization Report Number 01/2005*, Famine Early Warning Systems Network (FEWS NET), USA. Available from: <http://pdf.usaid.gov/pdf_docs/PNADH997.pdf> [accessed July 2012].

Giorgis, K., Tadege, A., and Tibebe, D. (2006) 'Estimating crop water use and simulating yield reduction for maize and sorghum in Adama and Miesso districts using the CROPWAT model', *CEEPA Discussion Paper No. 31*, Centre for Environmental Economics and Policy in Africa (CCEPA), University of Pretoria, Pretoria.

Goulden, M., Conway, D., and Persechino, A. (2009) 'Adaptation to climate change in international river basins in Africa', *Hydrological Science – Journal des Sciences Hydrologiques*, 54(5): 805–28 <http://dx.doi.org/10.1623/hysj.54.5.805> [accessed July 2012].

Grey, D. and Sadoff, C.W. (2007) 'Sink or swim? Water security for growth and development', *Water Policy*, 9(6): 545 <http://dx.doi.org/10.2166/wp.2007.021> [accessed July 2012].

Hedger, M. and Cacouris, J. (2008) *Separate Streams? Adapting Water Resources Management to Climate Change*, Tearfund. Available from: <www.preventionweb.net> [accessed July 2012].

Hurni, H.(1982) *Simen Mountains – Ethiopia Vol. II: Climate and the Dynamics of Altitudinal Belts from the Last Cold Period to the Present Day,* Geographisches Institut der Universität Bern, Bern.

Hurni, H. (1998) *Agroecological Belts of Ethiopia: Explanatory Notes on Three Maps at a Scale of 1:1,000,000,* Soil Conservation Research Program, Addis Ababa.

Intergovernmental Panel on Climate Change (IPCC) (2007) *Climate Change 2007: Impacts, Adaptation and Vulnerability. Contribution of Working Group II to the Fourth Assessment Report of the Intergovernmental Panel on Climate Change,* in M.L. Parry, O.F. Canziani, J.P. Palutikof, P.J. van der Linden, and C.E. Hanson (eds.), Cambridge University Press, Cambridge.

International Union for the Conservation of Nature (IUCN), Co-operative Programme on Water and Climate (CPWC) and World Water Council (2009) *Don't Stick your Head in the Sand! Towards a Framework for Climate-proofing, Perspectives on Water and Climate Change Adaptation,* IUCN, CPWC and World Water Council. Available from: <http://worldwatercouncil.org> [accessed July 2012].

Jones, L. and Boyd, E. (2011) 'Exploring social barriers to adaptation: insights from Western Nepal', *Global Environmental Change,* 21(4): 1262–74 <http://dx.doi.org/10.1016/j.gloenvcha.2003.10.006>.

Jones, L., Jaspers, S., Pavanello, S., Ludi, E., Slater, R., Arnall, A., Grist, N., and Mtisi, S. (2010a) 'Responding to a changing climate: exploring how disaster risk reduction, social protection and livelihoods approaches promote features of adaptive capacity', *ODI Working Paper 319,* ODI, London.

Jones, L., Ludi, E., and Levine, S. (2010b) 'Towards a characterization of adaptive capacity: a framework for analysing adaptive capacity at the local level', *ODI Background Note,* ODI, London.

Kaur, N., Getnet, M., Shimelis, B., Tesfaye, Z., Syoum, G., and Atnafu, E. (2010) 'Adapting to climate change in the water sector. Assessing the effectiveness of planned adaptation interventions in reducing local level vulnerability', *RiPPLE Working Paper 18,* RiPPLE, Addis Ababa.

Levine, S., Ludi, E., and Jones, L. (2011) 'Rethinking support for adaptive capacity to climate change: the role of development interventions: findings from Mozambique, Uganda and Ethiopia', *ACCRA Working Paper,* ODI, London.

Ludi, E., Jones, L., and Levine, S. (2011) *Preparing for the Future? Understanding the Influence of Development Interventions on Adaptive Capacity at Local Level in Ethiopia,* Africa Climate Change Resilience Alliance/ODI, London.

MacDonald, A., Davies, J., and Calow, R. (2008) 'African hydrogeology and rural water supply', in S. Adelana, and A. MacDonald (eds), *Applied Groundwater Studies in Africa* (IAH Selected Papers on Hydrogeology, 13) pp. 127–48, CRC Press/Balkema, Leiden.

Magrath, J. (2008) *Turning Up the Heat: Climate Change and Poverty in Uganda,* Oxfam GB, Oxford.

McCartney, M. and Smakhtin, V. (2010) *Water Storage in an Era of Climate Change: Addressing the Challenge of Increased Rainfall Variability,* International Water Management Institute (IWMI), Colombo.

McSweeney, C., New, M., and Lizcano, G. (2010a) *UNDP Climate Change Profiles: Ethiopia* [website]. Available from: <http://country-profiles.geog.ox.ac.uk> [accessed July 2012].

McSweeney, C., New, M., and Lizcano, G. (2010b) 'The UNDP climate change country profiles: improving the accessibility of observed and projected climate information for studies of climate change in developing countries', *Bulletin of the American Meteorological Society*, 91(2): 157–66 <http://dx.doi.org/10.1175/2009BAMS2826.1> [accessed July 2012].

Müller, C. (2009) *Climate Change Impact on Sub-Saharan Africa*, German Development Institute, Bonn.

Nicol, A. and Kaur, N. (2009) 'Adapting to climate change in the water sector', *ODI Background Note*, ODI, London.

National Meteorological Agency (NMA) (2007) *Climate Change National Adaptation Program of Action of Ethiopia*, NMA, Ministry of Water Resources, FDRE, Addis Ababa.

Nori, M. and Davies, J. (2006) *Change of Wind or Wind of Change: Climate Change, Adaptation and Pastoralism*, World Initiative for Sustainable Pastoralism, IUCN, Nairobi.

Oates, N., Conway, D., and Calow, R. (2011) 'The mainstreaming approach to climate change adaptation: insights from Ethiopia's water sector', *ODI Background Note*, ODI, London.

Organisation for Economic Cooperation and Development (OECD) (2009) *Integrating Climate Change Adaptation into Development Co-operation: Policy Guidance*, OECD, Paris. Available from: <www.oecd.org/dataoecd/0/9/43652123.pdf> [accessed July 2012].

Pelling, M. (2010) *Adaptation to Climate Change: From Resilience to Transformation* (1st ed.), Routledge, London.

Schipper, L.E.F. (2007a) 'PASDEP screening: assessing the entry points for integrating climate change into Ethiopia's development', in D. Conway, L. Schipper, M. Yesuf, M. Kassie, A. Persechino, and B. Kebede, *Reducing Vulnerability in Ethiopia: Addressing the Issues of Climate Change: Integration of Results from Phase I*, Overseas Development Group, University of East Anglia, Norwich.

Schipper, L.E.F. (2007b) 'Climate change institutions in Ethiopia: typology of actors, relationships and networks', in D. Conway, L. Schipper, M. Yesuf, M. Kassie, A. Persechino, B. Kebede, *Reducing Vulnerability in Ethiopia: Addressing the Issues of Climate Change: Integration of Results from Phase I*, Overseas Development Group, University of East Anglia, Norwich.

Tucker, J. and Yirgu, L. (2010) 'Small-scale irrigation in the Ethiopian highlands: what potential for poverty reduction and climate adaptation?' *RiPPLE Policy Brief 3*, RiPPLE, Addis Ababa.

United Nations Development Programme (UNDP) (2007) *Human Development Report 2007/2008. Fighting Climate Change: Human Solidarity in a Divided World*, Palgrave Macmillan, New York.

UNFCCC, *National Adaptation Programmes of Action (NAPAs) Online Databases* [website], UNFCCC, Available from: http://unfccc.int/national_reports/napa/items/2719.php [accessed July 2012].

Urama, K.C., Ozor, N. (2010) *Impacts of Climate Change on Water Resources in Africa: The Role of Adaptation*, African Technology Policy Studies

Network (ATPS), Nairobi. Available from: <www.ourplanet.com/climate-adaptation/Urama_Ozorv.pdf> [accessed July 2012].

World Bank (2006) *Ethiopia: Managing Water Resources to Maximise Sustainable Growth. A World Bank Water Resources Assistance Strategy for Ethiopia*, World Bank, Washington, D.C.

Water and Sanitation Programme (WSP) (2010) *Climate Risk Screening of the WSP Portfolio in India: Identifying Key Risk Areas and Potential Opportunities*, World Bank, Washington, D.C.

Ziervogel, G. and Zermoglio, F. (2009) 'Climate change scenarios and the development of adaptation strategies in Africa: challenges and opportunities', *Climate Research*, 40: 133–46 <http://dx.doi.org/10.3354/cr00804>.

About the authors

Lindsey Jones is a Research Officer at the ODI. His background is international development and environmental policy, having previously worked for the United Nations Development Programme (UNDP) and the World Food Programme (WFP). Lindsey's research interests are in the linkages between climate change, adaptation, and development. He holds an MSc in Environmental Policy from the University of Oxford and a BSc in environmental geography from the University of East Anglia.

Lorraine Coulter is a water specialist with expertise in the analysis of quantitative and qualitative livelihood systems information. She has a specific focus on the development of models to simulate and evaluate the impact of hazards on populations for use by developing country governments, United States Aid Agency (USAID) Famine Early Warning Systems Network (FEWS-NET), the World Bank, and various NGOs. Lorraine also leads emergency assessments and analyses to help decision-makers identify in-need populations and required support in times of drought, market shifts, and other shocks.

Million Getnet Gebreyes is a Lecturer in the Department of Rural Development and Agricultural Extension at Haramaya University. His research interests include participatory approaches, agricultural extension, climate change and pastoralism. Million holds an MSc in agricultural extension, and has led two multidisciplinary research teams in assessing the contributions of local development interventions towards adaptation in Ethiopia.

Beneberu Shimelis Feleke is a Senior Lecturer in the School of Natural Resource Management and Environmental Science at Haramaya University. Beneberu has worked for both national and international NGOs as a Research Consultant and Trainer. He has published numerous articles, book chapters, and working papers on the subjects of adaptation and adaptive capacity, and water management.

Naomi Oates is a Programme Officer at the Overseas Development Institute (ODI). Her background is in environment and international development, having previously worked on the Climate and Development Knowledge Network (CDKN). She holds an MSc in Climate Change and International Development from the University of East Anglia and a BSc in natural sciences from Durham University. Her research interests are climate change adaptation/resilience and natural resources management.

Leulseged Yirgu Gebreamlak is a RiPPLE Policy Officer and has contributed to various programme and research activities in relation to water and climate change. Leulseged holds an MSc in Regional and Local Development Studies from Addis Ababa University.

Josephine Tucker is a Research Fellow at the ODI with a background in ecology and water management and particular experience in Ethiopia and East Africa. Her research interests include water–food–ecosystem linkages, the role of water in resilience and disaster risk management (DRM), and equitable governance of water in rural and urban contexts. She is experienced in supporting multi-stakeholder platforms which link research, policy, and practice.

CHAPTER 8

Policy and practice influence through research: critical reflections on RiPPLE's approach

Josephine Tucker, Ewen Le Borgne and Marialivia Iotti

Research-inspired Policy and Practice Learning in Ethiopia and the Nile Region (RiPPLE) the programme did not follow a traditional research approach, but worked with Learning and Practice Alliances (LPAs) to carry out action research that responded to real problems on the ground. The aim was to ensure that research was relevant to sector needs, promote the uptake of findings into policy and practice, and build interest and capacity in the sector for the use of research and evidence-based approaches. This chapter describes RiPPLE's action research model, as well as complementary strategies for influencing and capacity building, and asks whether the programme achieved these goals. The chapter focuses on the first five years of the programme (2006–11), during which it received UK funding and systematic international backstopping. RiPPLE now continues life as an independent NGO, building on the same approaches, but its most recent activities and impacts are beyond our scope.

The context: a priority sector, but limited learning

Over the past decade, the Government of Ethiopia has prioritized water, sanitation, and hygiene (WASH) in national development policies. Funding has been mobilized for an ambitious Universal Access Plan (UAP), with progress made on harmonization and alignment around a national WASH fund (see Chapter 1). However, the planned pace of change may outstrip *woreda* (district) capacities to plan, monitor, and sustain services. Despite recent progress, the fragmented history of the sector (with many donors and NGOs providing funds), coupled with frequent changes in ministerial mandates and longstanding difficulties in coordinating stakeholders around integrated strategies, means that there has been little systematic documentation of the effectiveness of past and current approaches.

The absence of consistent procedures for storing and using data, and the resulting lack of institutional memory, leaves Ethiopia without a strong knowledge base for decision-makers or practitioners. Policies are rolled out from the top down, with limited feedback or innovation from the local level, while research in the sector has been piecemeal and has not always identified practical solutions to development challenges. This situation is exacerbated by limited interaction between academic and policy/practice communities (Box 8.1).

RiPPLE's early discussions with the Research Department of the Ministry of Water Resources (later the Ministry of Water and Energy – MoWE) identified a vicious cycle in which research was institutionally and financially de-prioritized, poorly linked to policy and practice, and failing to enhance social and economic development.

Box 8.1   The struggle for information

'Actors in the WASH sector find it very difficult even to find simple information … because information is scattered here and there'. (Coordinator, Millennium Water Programme)

During the development of RiPPLE from 2005–6, a wide range of stakeholders highlighted the challenge of accessing information and data for use in planning or implementing WASH services. Information tended to remain with individuals, and government and NGO staff reported phoning friends and contacts to obtain data, while the media were often unable to access information for public reporting. There was no repository for sector information other than a website managed by the Italian Development Cooperation on behalf of the European Union (EU) Water Initiative, and on-the-ground data collection and storage was patchy. *Woreda* water staff had little time or money for monitoring, and data collected were usually reported to higher government levels against targets, rather than maintained systematically at local level to inform planning. Meanwhile, research commissioned at national level rarely filtered down to *woredas* as concrete guidance. This disconnect meant that local plans were rarely evidence-based, but followed what one zonal office representative described as: 'the traditional trial and error approach of sitting in the office, planning and intervening'.

The new National WASH Inventory (NWI) will generate a wealth of new data about water access and scheme status, and has been supported and strengthened by RiPPLE. But questions remain over the quality of data, the form in which it will be made available to implementers and the public, and whether it will meet the needs of local planners as well as national actors focused on coverage statistics (see Chapter 2).

Source: quotes and text from RiPPLE reflective interviews with LPA and Forum for Learning on Water and Sanitation (FLoWS) members

---

Evolving theory on research–policy linkages

The disconnect between research and the development of policy and practice is not unique to Ethiopia. Conventional research often implicitly assumes a linear path from knowledge generation by experts to uptake by policy-makers and practitioners. Yet research rarely influences policy. Current thinking stresses that policy impact has several dimensions: attitude change, discursive commitments, procedural change, influence on policy content, and behaviour change (Jones and Villar, 2008; Keck and Sikkink, 1998). Identifying the desired change and understanding what drives change in target individuals or organizations are the first steps to policy influence. Various theories explain how policy change occurs, emphasizing the role of coalitions for advocacy, the need to seize 'windows of opportunity' for policy change, the importance of how messages are framed, and the need to engage decision-makers directly (Jones, 2011; Stachiowiak, 2007). The DFID (2008: 19) notes that the ability to access and use research is linked to the ability to *do* research, arguing for 'improving research capability by supporting environments that encourage people to use research'. RiPPLE drew on these theories to address common barriers to policy influence (Table 8.1).

Table 8.1 RiPPLE strategies for policy influence: tackling common challenges

Barriers that limit the policy influence of research[1]	RiPPLE strategies to maximize influence
Inadequate supply of relevant research	Produce new evidence responding to demand from policy-makers and practitioners
Poor or inequitable access to research, data, and analysis	Disseminate research widely; establish website as a repository; support resource centres
Lack of understanding of the policy process by researchers/ ineffective communication of research	Engage policy-makers from the start; monitor policy influence; establish communication and policy engagement team; validate emerging narratives with end-users
Lack of awareness among politicians, civil servants, and practitioners about available research	Establish fora to share research at regional and national level; support resource centres; unite academic and policy/practice communities
Anti-intellectualism in government (sometimes associated with blocking access to information for researchers)	Demonstrate the value of research by generating evidence that supports the government's mission; engage with the policy process on a continued basis over time
Poor capacity to absorb research (lack of culture, systems, time or resources to absorb research)	Provide practical support to information management systems and training in their use; offer training in research methods; involve end-users in research
Politicization of research, power relations, and contestation of what constitutes evidence	Maintain independence; share methodologies transparently; ensure research is led by experienced researchers; use interdisciplinary approaches
Disconnection between researchers and decision-makers, and/or between these groups and the wider public	Unite researchers and decision-makers in platforms and research teams; promote participatory methods; train media in WASH issues; disseminate findings through media

1 adapted from Stone et al., 2001

A new approach: action research driven by Learning and Practice Alliances

RiPPLE believed that research should respond directly to the needs of sector practitioners, policy-makers, and ultimately service users, and therefore adopted an 'action research' model (Box 8.2) in which:

- sector stakeholders identified research priorities and questions;
- end-users participated in the research alongside experienced researchers;
- research was carried out in partnership with implementers, and assessed the effectiveness of implementation on the ground;
- understanding the underlying causes of problems was paramount, with exploratory and participatory methods emphasized alongside quantitative diagnoses;
- the primary goal was to understand the implications of findings for policy and practice;
- the collaborative research process itself led to new observations, discussions, and insights.

Box 8.2 Action research

Action research engages potential research users (e.g. policy-makers, planners, and implementers) in a process of 'learning by doing'. Practical solutions to problems are developed as part of the research process. In contrast with traditional research, action research is linked to implementation and has an explicit agenda for change. It is often seen as a cycle in which a team of researchers and stakeholders jointly identify desired outcomes, diagnose problems, pilot approaches, evaluate their impact, propose improvements, and so on. The risks associated with involvement of non-researchers can be mitigated by the close involvement of experienced researchers to ensure data quality, while the quality of the *insights* generated is enhanced when findings are analysed by those who are familiar with realities on the ground.

Source: Tucker, 2008b

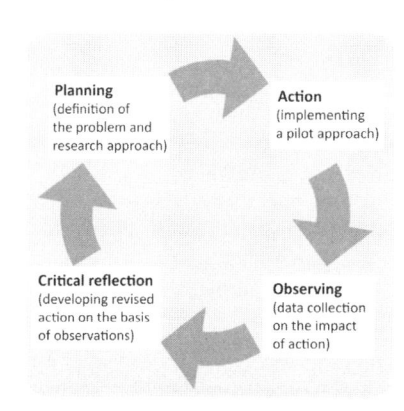

Planning (definition of the problem and research approach)

Action (implementing a pilot approach)

Observing (data collection on the impact of action)

Critical reflection (developing revised action on the basis of observations)

Cycles of action research

Source: adapted from Moriarty, n.d.

RiPPLE's action research was undertaken through LPAs – stakeholder platforms from *woreda* to national level – whose members set research questions, participated in research teams, and met regularly to discuss findings. Growing evidence suggests that working through LPAs enhances research uptake (Box 8.3).

> **Box 8.3 Learning and Practice Alliances (LPAs)**
>
> LPAs, or learning alliances, emerged in response to the frequent failure of conventional research to improve policy and practice. They bring stakeholders together to exchange knowledge, identify new solutions to problems, and develop joint agendas for change. Working with interconnected LPAs at multiple levels helps to scale up local innovations. Smits et al. (2007) and Moriarty et al. (2005) suggest that it is particularly important to work at intermediate levels, as these are vital to support community-level initiatives and can disseminate experiences and scale out innovations horizontally. The LPA approach requires sustained commitment and strong facilitation, especially in the early stages. It takes time for more collaborative ways of working to take hold, but this is the long-term goal.
>
> *Source:* Moriarty et al. (2005), Smits et al. (2007), and Tucker (2008b)
>
> **Some key success factors**
> Local enabling environment
> Adequate resources
> Good facilitation
> Good process documentation
> Time
>
> Partial exchange of information → Collaborative research → Systematic learning with common vision → Concerted action → Exerting influence of scale → Implementation at scale
>
> *Joint discussion* — *Joint work planning* — *Joint learning* — *Joint action* — *Joint advocacy and impact* — *Joint planning*
>
> **Improving collaboration and scale-up of innovation through LPAs**
> *Source:* Butterworth, 2006

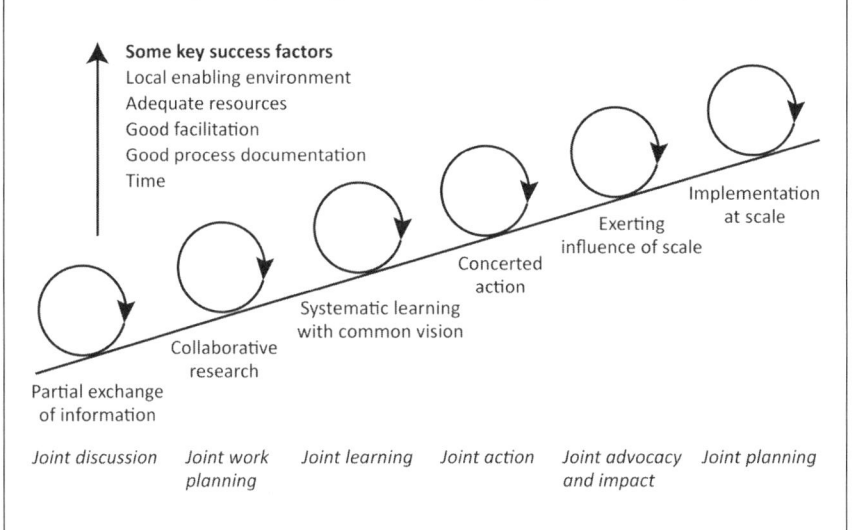

RiPPLE established LPAs in its research areas at regional/zonal level in the Southern Nations, Nationalities, and People's Region (SNNPR), Benishangul-Gumuz Regional State, and East Hararghe Zone in Oromia Region. LPAs were also established in two *woredas* in each of these regions/zones.

Regional and *woreda* LPAs are facilitated by full-time staff who arrange meetings, engage members in ongoing dialogue, and support them to act on research findings. LPAs are integrated through overlapping membership (each is attended by members from higher or lower administrative levels), cross-presentation of findings at meetings, and an LPA bulletin.

At national level the LPA Forum for Learning on Water and Sanitation (FLoWS) is hosted by the MoWE. Rather than focusing on RiPPLE's research, participants share work on selected themes and debate its implications. Themes have included multiple-use water services (MUS), the National WASH Inventory (NWI), climate change, sanitation, and sector financing.

As well as promoting uptake of RiPPLE research, LPAs aim to increase the use of evidence and promote information-sharing and coordination – i.e. to enhance learning.[1] Coordination between government and NGOs was limited at the outset, despite various national initiatives such as the multi-stakeholder forum (MSF). Joint planning between government offices was also weak: the Ministries of Water, Health and Education had committed to collaboration under a WASH Memorandum of Understanding (MoU) (see Chapter 1), but this had not been fully applied in practice, especially sub-nationally. Links between the water sector, agricultural policy, and drought preparedness were also limited. Figure 8.1 contrasts the LPA-driven research approach with a conventional research model assuming a linear path from evidence generation to uptake.

A conventional, linear transfer of technology model

Researchers generate evidence / develop innovations → **Policy-makers** adopt innovations in response to evidence → **Practitioners** implement new approaches for improved outcomes

A multi-stakeholder learning process

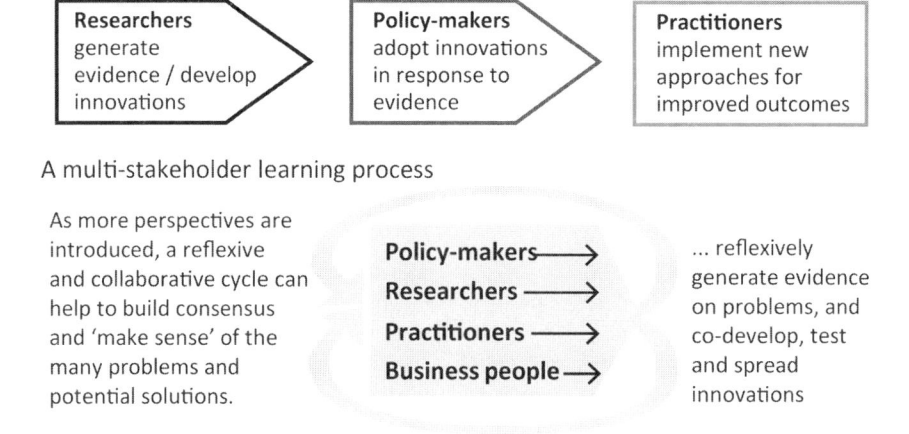

As more perspectives are introduced, a reflexive and collaborative cycle can help to build consensus and 'make sense' of the many problems and potential solutions.

Policy-makers →
Researchers →
Practitioners →
Business people →

... reflexively generate evidence on problems, and co-develop, test and spread innovations

Figure 8.1 Linear technology transfer model versus multi-stakeholder learning process

Source: Mason, 2011

Achieving wider influence

Participation in research or LPA meetings will not automatically lead to uptake of new evidence; it is important to translate findings into practical recommendations, and foster willingness and capacity of stakeholders to adopt new approaches. RiPPLE therefore undertook direct policy engagement activities and training, building on its research findings, usually working in partnership with others and through civil society coalitions (see Table 8.2).

Policy engagement strategy

Over its first two years (2006–8), RiPPLE shared diverse case study findings with policy-makers. Later research (2009–11) focused on two topics – extending and sustaining access through self-supply, and links between water, climate, and food security – with policy engagement targeted at:

- policies and tools for WASH, principally the accelerated Universal Access Plan (UAP) and National WASH Inventory (NWI);
- policies and strategies for climate change adaptation and food security, including the Growth and Transformation Plan (GTP), National Adaptation Programme of Action (NAPA), and Disaster Risk Management (DRM) strategy.

Figure 8.2 summarizes RiPPLE's research areas and policy engagement targets.

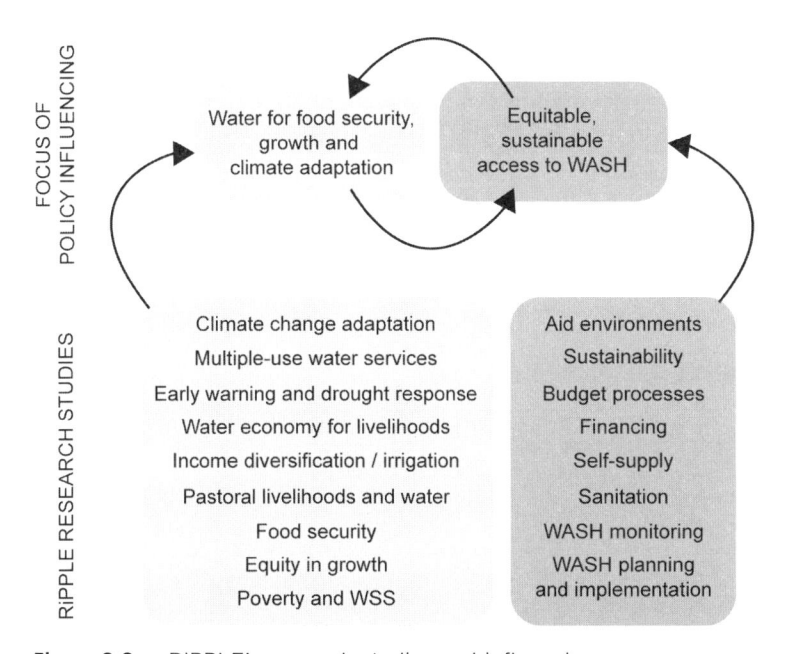

Figure 8.2 RiPPLE's research studies and influencing areas

Supporting improved practices

RiPPLE aimed to build the capacities of those who deliver and manage services at local level: community WASH committees (WASHCOs) and *woreda* water staff.

Following research identifying limited skills among WASHCOs as one cause of scheme non-functionality, RiPPLE developed training for WASHCOs and motor pump operators (piloted with over 400 WASHCO members and 85 pump operators in four *woredas)*. Operator training gave participants the chance to practise maintenance tasks, sometimes for the first time. Training was delivered jointly with *woreda* staff, improving relationships with WASHCOs and helping to standardize approaches between WASHCOs established by different implementers.

The second initiative focused on *woreda* staff. RiPPLE and partners SNV (Netherlands Development Organization) and MetaMeta developed and piloted a practical module for *woreda* water technicians and health staff, delivered by Technical and Vocational Education and Training Colleges (TVETCs): 'Guided Learning on Water and Sanitation' (GLoWS). GLoWS complements existing classroom-based teaching with on-the-job training to tackle real-life situations. Trainees follow distance-learning modules and carry out problem-solving exercises on the job, with regular visits from trainers. This culminates in the development of community water and sanitation plans covering: water supply improvement; water quality risk management; sanitation and hygiene; and technical and financial management of water points.

Public information and awareness

RiPPLE also seeks to raise wider awareness of WASH issues and policies, to promote public accountability of WASH policy-makers and implementers. In particular, RiPPLE produced a film highlighting the importance of safe and reliable water services, which was screened in the main square in Addis Ababa and at various public events, and sponsored a series of weekly programmes on national radio (with WaterAid and the WASH Movement). These featured RiPPLE's research findings, community interviews giving local perspectives on WASH, and dialogues with policy-makers.

RiPPLE's impacts and outcomes

RiPPLE is one of many actors in the sector and often works through networks and coalitions, in a context of rapidly evolving policy. This makes it hard to isolate RiPPLE's impact or attribute changes directly to the programme. Furthermore, the kind of change RiPPLE seeks is gradual, long-term, and not always easily measured (policy change is relatively easy to observe, but changes in mindsets or institutional culture are much harder to capture). Nonetheless, RiPPLE seeks to understand its impacts through regular reflective interviews with LPA and FLoWS members.[2] These provide some valuable evidence of the nature and extent of RiPPLE's influence, and lessons for similar programmes.

Table 8.2 Major RiPPLE influencing strategies beyond the LPA

Areas of engagement	Activities
Policy influencing	
National WASH inventory/ sector monitoring	National Symposium on WASH Monitoring, convened with the Consortium of Christian Relief and Development Associations (CCRDA) and International Water and Sanitation Centre (IRC)[1]
Linking water with food security and climate change adaptation	National- and zonal-level workshops with water, agriculture, and DRM officials Evidence contributed to Poverty Action Network Ethiopia (PANE) submission to GTP[1]
Innovative approaches to rural water supply	Symposium on multiple-use water services (MUS), convened with the MUS Group[1]
Practice influencing	
Developing practical skills	Training of WASHCOs and pump operators Training of *woreda* water staff in analysis of inventory data Training for journalists on WASH issues, with the Ethiopia Country Water Partnership (ECWP) and WaterAid[1] Support to Haramaya University to train practitioners on: participatory research methods, and water-related topics including climate change impacts and adaptation[1]
Strengthening curricula of universities and training institutions	Development and piloting of GLoWS, with MetaMeta and SNV[1] Development of university modules on WASH

1 undertaken in coalition with others

Uptake of findings into policy

Woreda- and regional-level uptake

Various new approaches have been adopted in focal regions/*woredas* following RiPPLE's research (Boxes 8.4 to 8.6). These show that evidence has strengthened the hand of *woreda* water offices in lobbying for more resources, and highlight the importance of including actors with budgetary responsibility in the research and LPA process. Change occurred when *woreda* administrators and regional bureaux were persuaded of the case.

Box 8.4 Enhancing scheme sustainability in two *woredas*

'We realized that frequent supervision improved how things were working. This had not been done frequently beforehand. We also realized the importance of coordinated action between the *kebele* (community) office and the water committee ... After we understood the problem it causes, we acted.' (Previous head of *woreda* water office, Mirab Abaya)

A RiPPLE study on scheme sustainability documented the high proportion of schemes that were non-functional, tracing the problems, in part, to lack of *woreda* resources. The *w*oreda water office in Halaba, SNNPR, used the findings to lobby for more funds. Its annual budget was increased eight-fold in 2009 – from 135,000 Ethiopian birr (US$7,500) to over ETB 1 million ($55,000)[3] – and the water office joined the *woreda* cabinet.

The Halaba Water Office also won over ETB 1.5 million ($83,000) from the regional bureau for two scheme rehabilitation and expansion projects, thanks to a combination of credible local evidence on functionality rates and increased understanding of the problem at regional level following discussions at the regional LPA. The water office also used evidence on scheme failure rates and levels of access to persuade three NGOs to invest in scheme rehabilitation and construction. Another result of the RiPPLE study was the decision, at regional level, to adopt submersible pumps rather than monopumps, as these were found to be more reliable.

In Mirab Abaya, SNNPR, the *woreda* water office responded to the findings by increasing its supervision of *kebele* offices and WASHCOs, following evidence that they needed more support.

Source: developed by the authors from an analysis of RiPPLE's impact using information from RiPPLE's monitoring activities throughout the programme

Box 8.5 Expanding water services for multiple uses

'When the LPA research started, multiple-use services (MUS) were only implemented by HCS [an NGO]. Now we know that, by adding some initial investment, we can implement MUS schemes, rather than single-use schemes, for great benefit of our communities. Our office is now using our World Bank budget to implement MUS schemes in the woreda.' (Vice Administrator, Goro Gutu *Woreda*, East Hararghe)

The Goro Gutu Water Office secured a doubling of its budget and new staff after a RiPPLE case study on MUS found that investing in MUS would be cost-effective in the long term compared with single-use approaches (see Chapter 3). As a result, multiple uses are now incorporated into the planning of all new schemes in the *woreda*, not just NGO schemes. The population with access to MUS more than tripled from 2005/6 to 2009/10, with over 1,400 households being served by 2010 and expected benefits in the order of ETB 800 ($44) per household per year (Adank et al., 2008).

Source: developed by the authors from an analysis of RiPPLE's impact using information from RiPPLE's monitoring activities throughout the programme

> **Box 8.6 Improving life for farmers: equitable water supply and better prices for vegetables**
>
> 'The study on income diversification showed us that lack of storage for vegetables is a problem. After a bumper crop, prices drop and communities do not want to sell at low prices, but have nowhere to store the vegetables. In collaboration with Haramaya University, I have now started to build a warehouse to store vegetables'. (Vice Administrator, Goro Gutu *Woreda*)
>
> In Goro Gutu, East Hararghe, a RiPPLE income diversification study on constraints to market access led to the construction of new vegetable storage facilities to enable farmers to store produce and sell in larger volumes or when prices are favourable. The private sector has also responded: research showing that vegetable prices were held down by a monopoly established by two local dealers drew new dealers to the area, and farmers now report that they receive better prices.
>
> In Mirab Abaya, documentation of inequities in access to irrigation by RiPPLE's equity study led to the establishment of *kebele*-level irrigation committees, new by-laws for equitable water allocation between better-off and poor farmers, and dispute resolution fora. The reported impact includes greater transparency in allocation of irrigation water, better record-keeping, and shorter water queues.
>
> *Source:* developed by the authors from an analysis of RiPPLE's impact using information from RiPPLE's monitoring activities throughout the programme

Although these changes occurred in only a few of Ethiopia's more than 700 *woredas*, they illustrate how credible evidence, combined with a collaborative process for stakeholders to discuss findings, can change ways of working from the bottom up. 'Credibility' means more than rigorous research methodology, although it was important that teams were guided by expert researchers. RiPPLE's research over the five years derived credibility from the transparent and inclusive way in which question-setting, data collection, and validation of findings were conducted. As one LPA member from a university put it: 'RiPPLE research works are credible due to the involvement of different stakeholders, preventing a few from dominating the research findings'.

It was also vital that research responded to the priorities of potential users, and helped them develop solutions. A research team member from a zonal office, explained: 'I see the value added of [RiPPLE's] research studies because they are practically oriented ... Unlike conventional research they are targeted at solving problems'.

National-level policy influence

Some studies that generated local impact have been taken up at higher levels, while others were geared towards national debates from the start and have achieved federal-level influence (Boxes 8.7 and 8.8).

Box 8.7 Influencing the Universal Access Plan and National WASH Inventory

The value attached to RiPPLE's work on WASH is reflected in the invitation from the MoWE to join the Advisory Committee of the Multi-Stakeholder Forum. RiPPLE's work on the design and implementation of MUS has had national influence: following discussions at FLoWS, awareness-raising by NGOs with experience of implementing MUS, and quantitative evidence of its benefits from RiPPLE, MUS was recognized in the updated UAP as an approach that could enhance service sustainability. RiPPLE also contributed by jointly convening a national symposium on MUS.

Similarly, RiPPLE's research on self-supply and WASH monitoring, and the convening of a national stakeholder workshop on monitoring, have resulted in the inclusion of self-supply in the revised NWI. As self-supply is a priority in the revised UAP, its inclusion in the NWI will make monitoring of progress against access targets more meaningful. RiPPLE is also supporting the reconciliation of monitoring approaches between the MoWE and the Joint Monitoring Programme (JMP – Chapter 2). While RiPPLE is not the only organization working on monitoring, it is one of the few to have worked closely with government staff at regional and *woreda* levels on the 'nuts and bolts' issues of data collection, analysis, presentation, and use, and to bring these experiences to bear in national debates. RiPPLE's membership of the National WASH Steering Committee was crucial to this influence on the NWI.

Source: developed by the authors from an analysis of RiPPLE's impact using information from RiPPLE's monitoring activities throughout the programme

---

Box 8.8 Engaging on food security, climate change, and disaster risk management strategies

RiPPLE produced evidence showing that water is central to food security and resilience, and that climate variability already affects the health and livelihoods of poor households through impacts on water resources (see Chapters 6 and 7). RiPPLE brought these findings to the attention of policy-makers responsible for DRM and early warning, and convened a national stakeholder event to discuss ways forward. Although national debates around DRM continue, and the policy impacts of RiPPLE's work are yet to be seen, RiPPLE has put water access on the table in national DRM debates and generated interest from the Ministries of Agriculture and Water and from donors, such as the United States Aid Agency for International Development (USAID), in incorporating water more systematically into vulnerability assessment and drought preparedness.

On climate change policy, RiPPLE has worked through alliances and coalitions to put key issues on the political agenda. It is hard to trace specific impacts, but RiPPLE has been invited to join the cross-ministry Working Group of Climate Change and Adaptation Planning under the Prime Minister's Office, and can therefore contribute insights from research to senior policy-makers.

> **Box 8.8 Continued**
>
> The ultimate impact of RiPPLE's work on DRM and climate change cannot yet be judged. However, RiPPLE's research findings are respected and welcomed by senior stakeholders; RiPPLE has been invited to join important policy-making fora and has brought stakeholders together to discuss findings and policy directions, sowing the seeds for greater collaboration.
>
> *Source:* developed by the authors from an analysis of RiPPLE's impact using information from RiPPLE's monitoring activities throughout the programme

Some RiPPLE studies have had more policy influence than others. Research on sector financing in particular seems to have had little impact so far, partly because the subject is highly politicized and difficult to influence. In addition, this research originated in Benishangul-Gumuz, where RiPPLE's activities had to be reduced due to political unrest.

Strengthening practice

Training of WASHCOs and *woreda* water staff has been popular and pilots have yielded clear benefits (Boxes 8.9 and 8.10). WASHCO training has been extended beyond the pilot *woredas* by Sodo Hosanna (a local NGO) and the Regional Government of SNNPR has directed *woredas* to consider WASHCO training for all schemes in their planning. The Ethiopian Government also intends to implement GLoWS nationwide through TVETCs, with support from the United Nations Children's Fund (UNICEF) and in collaboration with the Finnish-supported Community-led Water, Sanitation and Hygiene (COWASH) programme.

GLoWS and WASHCO training have also raised interest in training in pilot areas. According to the zonal water office in East Hararghe: 'the community is now asking to get training on the management of water schemes, and experts working in offices are also asking for training'. RiPPLE is now formally recommended as a capacity building provider in the national WASH Implementation Framework (WIF) (FDRE, 2011).

> **Box 8.9 Improving scheme sustainability by building WASHCO capacity**
>
> 'The training has had a significant impact in almost all kebeles ... In Kola Barena *kebele* the WASHCO became capable of doing maintenance activities on their own. After the training and after having started to save money, the committee covered the cost for spare parts and managed the repair of a handpump for themselves. In Yayike *kebele* the financial savings were about ETB 1,000 ($55) and they were reluctant to show this money. But after the training, the WASHCO covered O&M [operation and maintenance] costs for the engine and pipes in the order of ETB 6,500 ($360) by themselves'. (Community Mobilizer, WASHCO pilot *woreda*)

Box 8.9 Continued

Following training, WASHCOs appear to be better placed to maintain and repair schemes, both technically and financially. Financial collections by trained WASHCOs have increased by almost 300 per cent through better book-keeping, collection, and financial management and the increased trust placed in WASHCOs by water users (see Figure). Reporting of breakdowns to *woreda* offices has also increased, and 17 schemes have been repaired successfully by WASHCOs working with *woreda* offices since the training finished. WASHCO members report that they now know which problems they can solve themselves and when to request support from the *woreda*, and improved reporting systems have been established for water users, *kebeles,* and *woreda*s, improving WASHCO accountability.

The training was a chance for WASHCOs to share their experiences and engage with *woreda* staff. This helped *woreda* officials understand the issues faced by WASHCOs and helped to motivate WASHCOs that had felt unsupported by the *woreda*.

Source: developed by the authors from an analysis of RiPPLE's impact using information from RiPPLE's monitoring activities throughout the programme

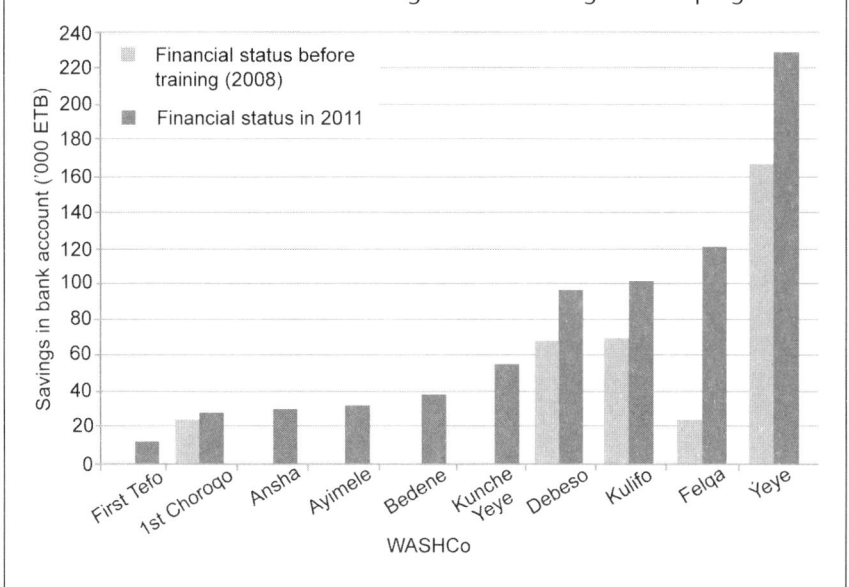

Money saved by WASHCOs before and after training (Halaba Special *woreda*)

> **Box 8.10 Practical planning skills for *woreda* water officers**
>
> Government staff who participated in the GLoWS pilot report benefits for themselves and the communities they serve. Practical exercises have improved their understanding of local conditions and given them the skills and confidence to respond to community needs. In particular, GLoWS helped local government staff engage communities in bottom-up planning processes whereby community members can articulate priorities and work with government staff on joint solutions. Trainees could use this approach, with support from experienced trainers, to solve practical problems.
>
> GLoWS also developed participants' ability to tackle management and institutional issues. Their achievements include: rehabilitation of shallow wells; cleaning of water points and motor pump sheds; replacement of a non-functioning WASHCO; reconnection of water supply to a school that had been disconnected for non-payment of bills; and scheduling separate access times or taps for community members and water vendors to reduce queuing times for girls. Coordination of sanitation promotion (with WASHCOs, health extension workers, school directors and *kebele* staff) has also improved.
>
> *Source:* developed by the authors from an analysis of RiPPLE's impact using information from RiPPLE's monitoring activities throughout the programme

Fostering evidence-based approaches

There is evidence that RiPPLE has encouraged the use of evidence; various LPA members indicate that research is now seen as useful to strengthen policies and practices in RiPPLE regions and *woredas*, and have noted increased demand for evidence at both *woreda* and zonal/regional level. Many had previously only conducted research at university, but now recognize that they can undertake, commission or use research to solve real problems. Their research capacity has been enhanced by membership of the LPA, and particularly by participation in research teams. Awareness and appreciation of participatory research methods in particular have increased. According to one respondent: 'there is little capacity [for research] in the sector, beyond what RiPPLE has created'.

Almost all LPA members surveyed in 2011 said that RiPPLE had inspired them to commission or undertake more research. According to an official from a regional health bureau: 'our office, including support staff in our *woreda* bureaux, is encouraged to do more research, having seen the achievements of the RiPPLE programme. It is helping us to fundamentally change the way we operate as regional government'. The East Hararghe Land Use and Environment Office has launched follow-up studies on adaptation options after a RiPPLE study showed what could be learned from speaking to farmers about climate change.

This reflects wider reports by stakeholders that RiPPLE's participatory approach has encouraged greater valuation of local knowledge and perspectives. Some government staff now actively seek the views of service users in planning; an irrigation engineer commented that: 'Being part of the LPA has increased my capacity in doing surveys on water supply and irrigation, assessing constraints faced by communities ..., research methodologies ..., team work ..., and participatory problem identification and prioritization, particularly involving end users in planning for better sustainability of the schemes'. One *woreda* office has assigned a dedicated staff member to collect information on WASH.

A donor partner, SNV, also reported that participation in research had encouraged their clients to undertake evidence-based planning, and civil society organizations (CSOs) have learned from RiPPLE's use of evidence to influence policy. One explained: 'now I have come to believe that we can influence decision-making at regional level if we provide strong evidence. This is a new way of thinking I have developed in the LPA'.

Stakeholders have observed a wider trend of increasing demand for evidence and data in Ethiopia's water sector. There is increasing recognition that better data and information systems would help the government monitor access, give donors a clearer baseline for programming, and allow local implementers to plan more effectively and develop funding proposals (see Chapter 2). In this context, RiPPLE's contributions have been well received, showing what can be achieved through collaborative, applied research at local level.

Finally, several of RiPPLE's university partners have increased their focus on WASH and incorporated participatory methods into their courses, following involvement in RiPPLE research (Box 8.11).

Box 8.11 Embedding research approaches in national universities

'The nature and purpose of the research created excitement [in the university] and we plan to be involved in similar types of activity in the future'. (Head of Department, Haramaya University)

There are signs that RiPPLE's partner universities have incorporated both RiPPLE themes and methods into their research and teaching. A researcher at Hawassa University (SNNPR) has replicated the sustainability case study in other *woredas*, while the Rural Development and Agricultural Extension Department of Haramaya University has included WASH in their curriculum. Several university lecturers involved in RiPPLE research also found it eye-opening to be exposed to qualitative and participatory methods that solve real-life problems (while their inputs in turn ensured academic rigour and introduced practitioners to more formal techniques), and have since incorporated these into their teaching and research. Participatory research methods are now being taught at Haramaya University, and the research training materials produced by RiPPLE on qualitative methods are part of the curriculum.

Source: developed by the authors from an analysis of RiPPLE's impact using information from RiPPLE's monitoring activities throughout the programme

Enhancing collaboration and learning in the sector

FLoWS and LPA members report growing interest in learning more broadly. FLoWS is valued for bringing experiences from different parts of the country, while implementers at zone and *woreda* level greatly value the insights gained by interaction with diverse stakeholders; one commented that it: 'enabled [him] to see things from different angles'.

This enthusiasm is illustrated by demands for RiPPLE to support resource centres (which it has done in two regions) and high rates of LPA attendance. One government representative observed that: 'the demand for information exchange platforms is evident ... [and] people are very attentive when they are called for such meetings'.

However, it is unclear how far this increased awareness extends beyond individual LPA members into institutional cultures and practices. Evidence on whether LPA member institutions have begun to collaborate outside LPAs is mixed. Some members report greater collaboration, thanks to contacts made during LPAs, and observe: 'an encouraging trend of connection outside the LPA platform'. Some felt that RiPPLE had helped to strengthen coordination between water, health, and education by providing regional fora and highlighting gaps in coordination at FLoWS. Others, however, questioned whether a non-governmental initiative could really drive intra-governmental coordination.

There has also been increasing coordination between government and NGOs in recent years. The national WASH Implementation Framework (WIF) states that 'the potential of CSOs/NGOs as capacity-builders is acknowledged and measures are being taken to ensure that their experience as innovators, grassroots implementers, and front-line trainers is effectively contributed to the wider WaSH programme' (FDRE, 2011: 111). Some attribute such shifts in part to FLoWS and the capacity building RiPPLE offered to NGOs; following a FLoWS meeting on the NWI, for example, NGOs were invited to join the NWI Taskforce.

Conclusion

LPA members agree that RiPPLE's most significant contribution to the sector was generating reliable local evidence. RiPPLE's independence from political agendas, and the participatory way in which the research agenda was set and results validated, were crucial to credibility and uptake, although some expressed frustration about the time this took. The diversity of LPA members was also important, as it ensured broad ownership of findings and created a critical political mass to drive change.

RiPPLE's multi-level approach was vital, bridging national policies, regional authorities, and local implementers. Regional bureaux were receptive to evidence-based proposals suggested by staff from *woreda* level, who had participated in research themselves, and regional support could therefore be leveraged to sustain and scale out local reforms.

A substantial communications effort and partnership with reputable organizations was vital initially to establish RiPPLE's profile in the sector, but influence was maintained thanks to high quality research findings and capacity building which met stakeholder demands. As an NGO partner commented: 'changes are observed because RiPPLE has picked up what is in line with the mission and interests of government and civil societies'.

Although there are many WASH actors in Ethiopia, RiPPLE targeted a gap by providing a national multi-stakeholder learning platform linked to local research, and gained influence by working closely and constructively with – but remaining independent from – the Ethiopian Government. Convening power at regional and *woreda* levels came from the selection of LPA facilitators who were well-connected with local stakeholders but again independent of government (mostly from well-known NGOs). Full-time facilitation by experienced local staff with dedicated offices was important to RiPPLE's successes, but was costly. It would be interesting to compare this with the impact achieved by similar platforms with less intensive facilitation.

Many stakeholders argued that RiPPLE's impact would have been greater if it had supported piloting of new approaches. But even without piloting, the combination of field exposure and discussions with new partners was valued by implementers: 'Even though there were a lot of water schemes constructed by NGOs in our *woreda*, I didn't get the chance to see them. During the research we did water scheme mapping and that was an unforgettable moment for me, to look at different water schemes and discuss issues with experts and different disciplines. I have gathered a large amount of information' (Engineer, Halaba Water Office).

These findings point to success on a key RiPPLE goal: building the confidence and capacity of *woreda* water offices to deliver services and coordinate the work of different actors. High-level support for capacity building for *woreda* staff beyond classroom-based training will be vital to sustain and scale up these local capacity gains beyond RiPPLE *woredas*, and the government's interest in scaling out GLoWS is a very positive outcome. However, a host of other actions are also required to increase motivation and professionalism at *woreda* level, including proper resourcing of *woreda* offices, and empowering and equipping staff to lead on local service development and management.

Bottom-up initiatives such as RiPPLE can help. A stronger evidence base on WASH access; the underlying causes of challenges; and the importance of WASH for broader development goals, combined with increased confidence and better performance at *woreda* level, can enable stronger advocacy for increased resources. RiPPLE's experience suggests that strengthening local-level capacities and evidence is particularly effective when combined with awareness-raising among those who control budgets and staffing.

However, the programme has had its limitations. It is hard to determine how far learning has gone beyond individual LPA members. This is a concern, given the high rates of staff turnover in government institutions and poor institutional management of information, although turnover may also spread new ideas rapidly across organizations. An Oromia Water Bureau staff representative commented: 'Being part of the LPA has enabled me to easily access all the research outputs and I use them personally. However, institutionally the research results are not well utilized except by some staff'. RiPPLE support for government resource centres is one attempt to overcome these problems. Many LPA members mentioned using research to inform further research or replication of studies, which is a positive sign and indicates that evidence is valued, but cannot be considered real uptake into policy or implementation.

RiPPLE also works in limited geographical areas. The programme has punched above its weight in some respects, having considerable impact at regional level in SNNPR for example, but substantial *woreda*-level practice changes look small from a national perspective. RiPPLE continues to conduct research and convene stakeholders around findings and policy issues, and the impact of these efforts needs to be tracked into the future. RiPPLE is in a strong position in the sector – respected and included in important policy fora and networks – but is now in a transitional phase with the withdrawal of systematic international backstopping, after UK funding stopped in June 2011.

Finally, the kind of changes that RiPPLE aims for in attitudes to learning and coordination are long-term. According to one informant: 'RiPPLE has injected the importance of learning into the sector', and there appears to be growing interest in the sector in capacity building, knowledge management, and collaboration. RiPPLE has supported this with robust evidence – the raw material for learning – but whether these gains are sustained depends on future sector politics.

Despite these limitations, some of them inevitable for a programme of this size and duration, we believe that RiPPLE's approach has proven its value by making research relevant and useable for actors in the WASH sector and beyond, achieving considerable influence at both local and national level, and starting to shift attitudes in favour of learning and the use of evidence. It is a concrete step towards the use of the research process itself as a way to transform the lives of ordinary citizens for the better.

Notes

1. Learning is magnified when it goes beyond the individual, as people interact and groups change their collective understanding of a problem and the actions needed to address it. Such 'social learning' occurs when change becomes embedded in wider social units and communities of practice through social interactions and networks (Reed et al., 2010). Social learning provides a broader knowledge base to address complex challenges, and helps to build a common vision and understanding among stakeholders leading to a greater chance of coordinated action.
2. All quotes are from these monitoring interviews, unless otherwise attributed.
3. Exchange rate US$1 = ETB 18.02 (17 July 2012)

References

Adank, M., Jeths, M., Belete, B., Chaka, S., Lema, Z., Tamiru, D., and Abebe Z. (2008) 'The costs and benefits of multiple uses of water: the case of Gorogutu *Woreda* of East Hararghe zone, Oromiya Regional State, eastern Ethioipa', *RiPPLE Working Paper 7*, Research-inspired Policy and Practice Learning in Ethiopia and the Nile Region (RiPPLE), Addis Ababa. All RiPPLE papers available from: <www.rippleethiopia.org/> [accessed July 2012].

Butterworth, J. (2006) *An Introduction to Learning Alliances for Scaling up Impacts of Research in IUWM*. Unpublished presentation to SWITCH Project Kick-off Meeting, IRC International Water and Sanitation Centre, The Hague.

Department for International Development (DFID) (2008) *Research Strategy 2008–2013*, DFID, London. Available from: <www.dfid.gov.uk/Documents/publications1/research-strategy-08.pdf> [accessed July 2012].

Federal Democratic Republic of Ethiopia (2011) *The WASH Implementation Framework (WIF) – Summary*, version: 27 July 2011, FDRE, Addis Ababa.

Jones, H. (2011) 'A guide to monitoring and evaluating policy influence', *ODI Background Note*, Overseas Development Institute (ODI), London.

Jones, N. and Villar, E. (2008) 'Situating children in international development policy: challenges involved in successful evidence-informed policy influencing', *Evidence and Policy* 4(1): 53–73 <http://dx.doi.org/10.1332/174426408783477891>.

Keck, M. and Sikkink, K. (1998) *Activists Beyond Borders: Advocacy Networks in International Politics*. Cornell University Press, New York.

Mason, N. (2011) *Transdisciplinary Research Protocol. A Guidance Manual for EAU4Food Researchers. Internal Programme Document*, EAU4Food Programme, Wageningen University and Research Centre, Wageningen.

Moriarty, P. (no date) 'A brief introduction to action research concepts and practice', *Learning Alliance Briefing Note 4*, SWITCH Project, IRC, The Netherlands.

Moriarty, P., Fonseca, C., Smits, S., and Schouten, T. (2005) *Background Paper for the Symposium on Learning Alliances for Scaling up Innovative Approaches in the Water and Sanitation Sector*, IRC, The Netherlands.

Reed, M. S., Evely, A. C., Cundill, G., Fazey, I., Glass, J., Laing, A., Newig, J., Parrish, B., Prell, C., Raymond, C., and Stringer, L.C. (2010) 'What is social learning?' *Ecology and Society* 15(4): r1. Available from: <www.ecologyandsociety.org/vol15/iss4/resp1/> [accessed August 2012].

Smits, S., Moriarty, P., Fonseca, C., and Schouten, T. (2007) 'Scaling up innovations through learning alliances: an introduction to the approach', in S. Smits, P. Moriarty, and C. Sijbesma, *Learning Alliances: Scaling up Innovations in Water, Sanitation and Hygiene*, IRC, Delft.

Stachiowak, S. (2007) *Pathways for Change: 6 Theories about How Policy Change Happens*, Organisational Research Services, Seattle.

Stone, D. (2001) 'Getting research into policy?' Third Annual Global Development Network Conference: Blending Local and Global Knowledge, 10 December 2001, Rio de Janeiro.

Tucker, J. (2008a) Action Research. *RiPPLE Information Sheet 3*, RiPPLE, Addis Ababa, Ethiopia.

Tucker, J. (2008b) Learning and Practice Alliances. *RiPPLE Information Sheet 4*, RiPPLE, Addis Ababa.

About the authors

Josephine Tucker is a Research Fellow at the Overseas Development Institute (ODI) with a background in ecology and water management and particular experience in Ethiopia and East Africa. Her research interests include water–food–ecosystem linkages, the role of water in resilience and DRM, and equitable governance of water in rural and urban contexts. She is experienced in supporting multi-stakeholder platforms which link research, policy, and practice.

Ewen Le Borgne is a Knowledge Sharing and Communication Specialist at the International Livestock Research Institute (ILRI) in Addis Ababa. At the time of writing, he was Programme Officer at the International Water and Sanitation Centre (IRC). He has worked on learning, knowledge management, monitoring and evaluation, and communications, in Ethiopia, west and east Africa, and Latin America. Ewen has extensive experience supporting resource centre networks, learning alliances and other multi-stakeholder 'sector learning' initiatives in the WASH and agricultural sectors, and has written papers related to knowledge management and learning in the WASH sector. He is a senior editor of the *Knowledge Management for Development (KM4Dev) Journal* and blogs on agile knowledge management for social change and empowerment at http://km4meu.wordpress.com. Ewen was co-lead for RiPPLE's communications, and monitoring and evaluation teams.

Marialivia Iotti is a Programme Manager at the ODI with more than 10 years' experience in managing complex, multi-year and partnership-based research programmes. Her background is in political science with particular experience in Ethiopia and China. Her interests include M&E, communications, and multi-stakeholder platforms, such as the learning platforms used in RiPPLE.

Index

action research 176box
adaptation: adaptive capacity 156–66; and food security 181t; necessity for 6; and pastoralism 19; planned 19, 158–63; and vulnerability 165 see also NAPA
advocacy 40
AfDB (African Development Bank) 32
African Minister's Council on Water see AMCOW
agriculture: cash crops 137; double cropping 30; as employer 133; importance of 30; intensification 135; low prices for produce 137; protection from climate change 164; water investment 133–5; yield increases 136
Alaba Special *Woreda* 14, 107, 111–12, 115, 116
alignment, reluctance to commit to 40
Alma-Ata, Declaration of 12, 89
AMCOW (African Minister's Council on Water) 33, 34
aquifer productivity 118fig
arsenic contamination15, 117
Asia 135
asset rebuilding 6
Australia 154

book-keeping 115
boreholes 15, 71, 112, 117, 118, 119, 154
BoWEs (Bureaus of Water and Energy) 44
budget utilisation rates 33
Burkina Faso 98, 99

capacity: government 111; lack of 36, 37,114; local 3, 6, 7, 41, 72, 111, 161–2; MoWE 40; RiPPLE 114; WASHCOs 36, 42–3, 44, 72, 116; *woreda* level 10, 32, 34box, 44, 58, 61–2, 63

capital investment 5, 6, 42, 75, 82, 107, 111
carbon trading 5
Care Ethiopia 140
CCRDA (Consortium of Christian Relief and Development Associations) 40, 181t
CDF (Community Development Fund) 8, 40–2, 116
census, national 53
Central Statistical Agency see CSA
child mortality 14, 16, 91, 129
China 2, 97
CHPs (community health promoters): capacity building 13; communication 96; equipment 96; motivation 13, 96, 101; non-monetary incentives 99; training 95, 96; transport 96; as voluntary staff 12, 92
CHWs (community health workers): attrition rates 99; definition, WHO 98; generalist role 100; importance of 13; and rural communities 98
climate change: Africa, regional impacts 147; and development 165–6; effect on rainfall 27; Ethiopia 150–2; global impact 147, 148box; lack of reliable data 4, 150; rainfall impact 148; regional assessment of vulnerability 19; response to 156–7, 159; RiPPLE research 184box; seasonal challenges 111; and sustainability 119box; threats to water security 5; vulnerability 6
Climate-resilient Green Economy see CRGE
CLTSH (Community-led Total Sanitation and Health) 91, 93, 100, 101, 118
CMPs (community managed projects) 69–85; CDF 41; finance 116; and functionality 71, 117;

priority for sustainability 108; scaling up of other schemes 116; and self-supply 82; *woreda* managed projects 70
collaboration, inter-agency 20
Community Development Fund *see* CDF
community health promoters *see* CHPs
community health workers *see* CHWs
community managed projects *see* CMP
Community-led Total Sanitation and Health *see* CLTSH
Consolidated WASH Account *see* CWA
Consortium of Christian Relief and Development Associations *see* CCRDA
contamination, well water 15, 26, 81, 91, 117
CRGE (Climate-resilient Green Economy) 5, 18, 19, 158
cross-subsidies 116
CSA (Central Statistical Agency) 9, 52
CWA (Consolidated WASH Account) 8, 32, 39

DAG (Development Assistance Group) 37, 38
DALYs (disability-adjusted life years) 16, 129, 130
data: accessibility of 60; collection, regional and local 58–60; local use of 60box; storage of 173
decentralization: effect of funding 33; health service 100; importance of 40; and lack of capacity 7; MUS 78; training 43; and UAP 25; water storage 164; *woreda* level 32, 58
deforestation 135, 157, 159
demand-responsive approaches *see* DRAs
Demographic and Health Surveys *see* DHS
'design capacity' 54
development, importance of water for 128
Development Assistance Group *see* DAG
DFID (UK Department for International Development) 32, 39

DHS (Demographic and Health Survey) 52
diarrhoea: DALYs 16, 129; financial cost of 16, 129; prevention of 12, 89, 98, 131–3; RiPPLE study 112; and seasonal water access 134fig
disability-adjusted life years *see* DALYs
disaster risk management *see* DRM
disaster risk reduction *see* DRR
disease: cost of treatment 133; and livestock movement 140; prevention 11; safe water 73; tropical 89; water related, by age group 131t *see also* diarrhoea; HIV/AIDS; malaria; parasites, intestinal; TB
diversification, income 4, 153, 183box
documentation, importance of 21
'domestic plus' entry point MUS 74, 75, 75t, 76fig
DRAs (demand-responsive approaches) 110
DRM (disaster risk management) 184–5box
drought 4, 5, 18, 27, 164
DRR (disaster risk reduction) 6, 25, 165
dysentery 131–3

East Harghe 11, 33, 34box, 131fig, 185
economic growth, and water access 130, 153
ecosystems 31, 135
education, children's 34box, 130, 138
El Niño 149
employment, off-farm 16, 133
emergency relief, short-term 140
enforcement, sanctions 95, 96
environmental factors, non-functionality 112
EPA (Environmental Protection Authority) 19, 158, 166
ESRDF (Ethiopia Social Relief, Rehabilitation and Development Fund) 32
Ethiopia: agricultural production as low 164; changing climate 18, 150–2; climate profile 149–50; distribution of water 25–7; geography of 26; government policy on water access 7; growth potential 128;

INDEX 197

land suitable irrigation 30; mitigation of climate change 159; population distribution 27; and predominantly rural population 2; rainfall 135; rainfall regions 150fig; river basins 27fig; rural access to water 7, 51, 54, 70; surface water per capita 135; urban water supply coverage 51t; water 129–33
Ethiopian Water Technology Centre 44
EU (European Union) 38, 174box
evapotranspiration 119box, 162
excreta disposal 12, 91, 92

FAO (Food and Agriculture Organization) 17
federal government: Basin High Council 8; cooperation 9; MoU 37, 43box; NWI 56, 58; WASH 33, 55; support as inadequate 36; water supply coverage 52box
floods 4, 5, 18, 27, 164
FLoWS (Forum for Learning on Water and Sanitation) 20, 189
fluoride contamination 15, 26, 117
Food and Agriculture Organization see FAO
food security 6, 19, 73, 74, 127–8, 130, 133, 184box
foreign investment 40, 135
Forum for Learning on Water and Sanitation see FLoWS
funding: agriculture 164; government grants 41; inadequate, and water shortage 127, 154; increases in 71; irrigation 165; MoFED 39; productive use of 6; public 33

GCMs (Global Circulation Models) 151–2
GDP (gross domestic product) 18, 128, 153
gender see women
geographical coverage, inconsistency in 13
GHG (greenhouse gas) emissions 17, 148
Global Circulation Models see GCMs
GLoWS (Guided Learning on Water and Sanitation) 3, 6, 15, 101, 185, 187box

GPS (Global Positioning System) 58box
Green Economy Strategy 135
greenhouse gas emissions see GHG
gross domestic product see GDP
groundwater: boreholes 112; and climate change 18; contamination 118; dependence upon 4; development of 163, 164; effect of climate change on 119; falls in levels 15; impact of rainfall variability 154; influences upon 26; irrigation 136; resources 25; rural reliance upon 2, 117; storage, Africa 120fig; wells 30, 78
GTP (Growth and Transformation Plan) 7, 17, 25, 29–30, 80, 135
Guided Learning on Water and Sanitation see GLoWS

Halaba Special *woreda* 94, 96
handpumps 3, 14, 72, 81, 107
handwashing 12, 16, 92, 94, 132
hardware 12, 34, 90, 93
health education 89, 90
Health Extension Programme see HEP
health extension workers see HEWs
healthcare costs 16, 132
HEP (Health Extension Programme) 89–102; curative 93; government grants 42; government support needed 97; parallel to government 14; philosophy 93; prevention 93, 100; RiPPLE study 11; S&H 92t; sanitation awareness 37
HEWs (health extension workers) 12, 13, 92, 93, 96, 100
HIV/AIDS 92t, 98, 99, 100, 129
household surveys 50, 55
hydropower 5, 8, 19, 30, 128, 159
hygiene education 12, 82, 92, 93, 94, 97

Ido Jalala 74–8
Ifa Daba 74–8
IFAD (International Fund for Agricultural Development) 136, 140
IMR (infant mortality rate) 91
India 2, 115
industrial activities, small scale 8, 73
inequality, of wealth 155
infant mortality rate see IMR

198 ACHIEVING WATER SECURITY

information, unreliability of 8, 9, 28, 50, 55, 110, 163, 174box
institutional arrangements: exclusion from planning stage 9; importance of 19; lack of 153; and non-functionality 72
Integrated Water Resources Management *see* IWRM
integration, water and other services 6, 38, 43
International Drinking Water Supply and Sanitation Decade 50, 69
International Fund for Agricultural Development *see* IFAD
IPCC (Intergovernmental Panel on Climate Change) 18, 148box
irrigation: crop production 83; difficulty of extension 78; effect on GDP 136; effects mitigated in Africa 135; Ethiopian potential 17; expansion sought GTP 30; income increase 137; investment in 8; irrigation canals 75, 76; 'irrigation plus' entry point, MUS 74, 75t, 76fig; management, as essential 136; mitigation of rainfall variability 19; need to expand coverage 165; potential 136; and poverty alleviation 136; reservoirs 75; small scale 135, 136; weak institutions and management 136
IWRM (Integrated Water Resources Management) 31

JMP (Joint Monitoring Programme) 4, 7, 8, 9, 28, 52–3, 110
JTR (joint technical review) 39

kebele: importance of 12, 95, 182box; irrigation committees 183box; and NWI 57, 58, 61, 62; and self-supply 42; WASH Teams 97

La Niña 149
land ownership 130
latrines 12, 37, 89–94, 99, 132
literacy 80
livestock management 17, 30, 83, 128, 138, 140, 141, 156fig
LPA (Learning and Practice Alliance) 20, 94, 173, 176–8, 187–8, 189

M&E (monitoring and evaluation): differences in results 54–5, 56; difficulties of 9; improvements in 37; limitations of 3; sector-wide 39; WASH outcomes 6; weakness of 7, 107
malaria 92t, 99, 100, 129, 131
malnutrition 16, 129
managerial capacity, lack of 70
marginalization 157
MDG (Millennium Development Goal) 2, 28, 49, 50, 70, 129
Memorandum of Understanding *see* MoU
messaging 12, 13, 91, 95–7, 99, 175
microfinance institutions 114
milk production 138
Millennium Declaration 8
Millennium Development Goal *see* MDG
Millennium Water Alliance 40
Ministry of Finance and Economic Development *see* MoFED
Ministry of Health *see* MoH
Ministry of Water and Energy *see* MoWE
Mirab Abaya *Woreda* 14, 71, 83, 84box, 94, 96, 107, 111–12, 114–15, 116, 118
mobile technology 115
MoFED (Ministry of Finance and Economic Development) 3, 32, 37, 38, 39, 41
MoH (Ministry of Health) 11, 37, 38, 90, 91
monitoring and evaluation *see* M&E
morbidity 38
mortality, premature 130
motivation, household 95
MoU (Memorandum of Understanding) 37, 38, 178
MoWE (Ministry of Water and Energy): capacity building fund 44; policy reform 7; responsibilities under WASH MoU 38; water supply responsibility 37; *woreda* and data 61
MSF (multi-stakeholder forum) 39, 178
MUS (multiple-use water services) 69–85; challenges to 78; community managed 74; costs 117; finance 116; integration 73; local 161; mainstream 11; 'MUS by design' 74; in pastoral areas 140; pathways 75fig; promotion of 73; self-supply 83–4

INDEX 199

NAPA (National Adaptation Programme of Action) 19, 31, 135, 158–9, 179
National Hygiene and Sanitation Strategy 93
National Sanitation and Hygiene Task Force 100
National WASH Inventory *see* NWI
National Water Resources Management Policy 42
NGOs (non-governmental organizations): data and collaboration 60; lack of coordination with government 178; management of projects 41; and NWI 58; partnership 40; and WASH 13, 32
Nigeria 28
non-functionality 71, 72, 107, 112–13
NWI (National WASH Inventory) 55–8; data 174box; data collected in 56, 57box; data collection methods validated 9; data use regionally and locally 60–1; donor funding and data 60; household surveys 82; and improved data reliability 37; and information provision 6, 101; as monitoring tool 9, 32; non-functionality figures 72; ownership of data 61; rural water supply 58box; scaling up 3; sector-wide 39

O&M (operation and maintenance): collection of costs 115–16; community 110; community training 69; household responsibility 73; responsibility of users 70; skills lacking 72, 111; subsidization 42
ODA (overseas development assistance) 2, 3
OECD (Organisation for Economic Co-operation and Development) 2
OMSUs (operation and maintenance support units) 16, 111, 115
open defecation 28, 91, 118
operational budgets 3
Organisation for Economic Co-operation and Development *see* OECD
Oromia Growth Corridor 141

Oromia Region 16, 79box, 80, 81, 83, 117
overgrazing 157
overseas development assistance *see* ODA

parasites, intestinal 112
Paris Declaration 49
pastoralism 17, 19, 71, 91, 140–1
PCDP (Pastoral Community Development Project) 140
personal hygiene 12
planning, long term 157
policy alignment 39
policy engagement strategy 179
population growth 2, 28, 111, 135, 153
poverty alleviation: and irrigation 135–8; as potentially lost 14; poverty reduction strategies 29–32; and vulnerability to climate change 5, 6; and water and sanitation extension 5; and water security 2, 16, 31, 130, 131fig
precipitation 152box, 50fig
Prime Minister, Ethiopia 159
privacy 13
private-sector development 40, 72, 99, 117
PSNP (Productive Safety Net Programme) 34box, 83, 133, 161, 165
Public Health Proclamation 2004 95
public information 180
public works programmes 6
pumping, irrigation 136
pumps 14, 78, 84, 112

rangeland management 161
regional government: campaign 12
research, and policy 174–5
reservoir capacity 153, 154
RiPPLE: and decentralization 164; evidence-based encouragement of 187–8; findings re irrigation 136–8; and HEP 89; importance of 5, 9, 21; importance of data re Ethiopia 129; influence 11, 179–80; influences upon 181t; on irrigation 165; local planning as essential 160; and local use of data 60box; MUS interventions 74; national policy influence

183–5; as NGO 173; and NWI 58; objectives 1; and policy influence 175t; predicted water coverage SNNPR 71; research in universities 188box; research on non-functionality 36; research on pastoral areas 140; research SNNPR 13; S&H 93; strengthening practice 185–7; study water stress 154; support for improved practices 180; survey of family wells 81; uptake of findings into policy 181–3
risk reduction 6, 25, 137, 165
River Basin Organisations 8
rope pumps 81, 84
Rwanda 100
RWSN (Rural Water Supply Network) 72

S&H (sanitation and hygiene): communication 95; education as limited 91; funding 101–2; HEWs and CHPs 96–7; regional variations in access 90
sanitation: access 14, 27–9; global access 2; CAPEX 35t; coverage 35t; definition 91; effect of improvements 130; improved, definition 91; institutional 36, 42; lack of subsidy 42; less progress than water services 7; population access to 90–2; rural population 37; sustainability 37
'sanitation ladder' 12, 94fig, 99
Save the Children USA 140
sector-wide approach *see* SWaP
sedentarization 140, 141
self-supply 69–85; CDF 42; definition 79box; encourages MUS 80; finance 116; government role 78; government support 108; important milestones 79box; as inexpensive 78; lack of strategy 82; promotion of 11, 73; scaling up 80–4; UAP 31
service options 6
sewerage, investment 36
skin conditions 131
small-scale irrigation 137, 160, 161, 165
smart-phone technologies 10, 115
SNNPR (Southern Nations, Nationalities, and People's Region): latrine construction 37; RiPPLE 111; S&H 91, 93–6; *Woreda* Inventory Survey 59, 60box
software 90
soil conservation 17
soil erosion 17, 135, 140
South Asia 19
Southern Nations, Nationalities, and People's Region *see* SNNPR
springs 112
standpipe, domestic 76
State of the World's Land and Water Resources (FAO) 17
Strategic Action Plan 91, 93
sullage disposal 91
support, lack of, post-construction 70
sustainability 36–7; change in attitude necessary 5; conceptual framework 109fig; definition 108; drivers of 108, 110; environmental challenges 117–20; as essential 4; financial challenges 115–17; institutional challenges 114–15; MUS 10, 11; policy and practice 110–11; and private sector 111; of progress, as difficult 7; social challenges 113–14; technical challenges 112–13; water services, Ethiopia 107–22
'sustainable intensification' 17
SWAp (sector-wide approach) agenda 38
systems alignment 39

tariffs 42, 116
TB (tuberculosis) 92t, 99, 100
Technical and Vocational Education and Training College *see* TVETC
technical capacity, lack of 70, 111
technologies, low cost 3, 4, 11, 33, 73, 108
technologies, protection, for drinking water 80
technology model 178fig
temperature increase, climate change 150, 151box
Thailand 97
tropical diseases 12
tuberculosis *see* TB
TVETC (Technical and Vocational Education and Training College) 3, 21, 43

INDEX 201

UAP (Universal Access Plan): achievements of 7; aims of 7, 25; definition minimum rural standards 52box; MUS 184box; private-sector development 117; revised 70, 72, 83; self-supply 79box; service level, water 36; success 107, 108; targets 28, 32, 33
UK Department for International Development *see* DFID
UN agencies 38
UNFCCC (United Nations Framework Convention on Climate Change) 158
UNICEF (United Nations Childrens' Fund) 8, 32, 56, 81, 110; JMP 50
United Nations *see* UN
United Nations Framework Convention on Climate Change *see* UNFCCC
universal access, as challenge 71
Universal Access Plan *see* UAP
USAID (United States Agency for International Development) 52

violence 29
VLOM (village-level operation and maintenance) 69
voluntarism 70
vulnerability alleviation 5, 140

WASH (water, sanitation and hygiene): alignment 39–40; capacity problems 34box; coordination 97; cost recovery 42; decentralization 40–2, 41; difficulties re institutional weaknesses, Ethiopia 3; and effect on other MDGs 2; extension to rural population 7; funding of 32–6, 41–2; harmonization 39; integration, water and other services 37–9; low priority public spending SSA 3; Management Information System 60; monitoring 49–64; motivation 114; partnership 40; pilot study 56; principles for implementation 43box; and public information 180; responsibilities 37; rural 2; unreliability of data 36; water supply and budget 34box; *woreda l*evel 34box WASH implementation framework *see* WIF

WASHCOs (water, sanitation and hygiene committees): capacity building 72; 185–6box; capital management 41; collection of revenue 115; community led 10; exclusion of women 15; lack of accountability 114; lack of capacity 72; low level capacity 36; O&M 70; and sustainability 42–4; training 15, 185
water: access, rural 4, 26, 34, 36t, 53t, 154; audits 14; CAPEX 35t; causes of scarcity 1; and climate change 162; Collection 113, 129, 133, 154; conservation 17; and disease 128; distance from access 36; distribution of 18, 25, 35t; drinking water access 28, 29fig, 78; institutional needs 36; quality control 12; reduction in ODA 2, 3; safe storage of 12, 92, 153, 164; seasonal access 134fig ; security as vital 2, 3; service definition 109; source types 77fig ; surface flow 25; tariffs 42; time saving 76, 129; urban 27–9, 42; variability 26–7, 154
Water Action 113
Water and Sanitation Program *see* WSP
Water Economy for Livelihoods *see* WELS
Water Extension Workers 116
Water Resources Management Policy 25, 31
water, sanitation and hygiene *see* WASH
water, sanitation and hygiene committees *see* WASHCOs
'water scarcity threshold' 26
Water Sector Strategy 25
WaterAid 60box, 180
watershed management 135
water-tables 119
Welfare Monitoring Surveys *see* WMSs
wells: contamination 81, 82, 118; family 11, 31, 54, 77–85; hand dug 5, 30, 71, 111; hand pumped 112; machine dug 112; protected 81; seasonality 117; sustainability 116, 118; traditional 15, 81; water quality 81
WELS (Water Economy for Livelihoods) 14, 110, 154

WHO (World Health Organization) 8, 50, 110, 130
WHO JMP 50, 71, 90
WIF (WASH Implementation Framework): addressing problems re organization 3; capacity building 185; harmonization as aim 39; implementation of self-supply 79box; KWT 97; objectives 99; principles 37; and private sector 16, 111; support for 32
WMS (Welfare Monitoring Survey) 16, 52, 130
Wolliso National Consultative Self-supply workshop 79box
women: as excluded from WASHCOs 15; as heads of household 137; marginalization of 112, 113, 157; participation in 110; water collection 29, 76, 128
woreda: analysis 130; capacity building 43, 44, 72; devolution to 32, 41; funding at level 32; importance of 58; managed projects 6, 41; motivation low 44; support for WASHCOs 43; technical expertise needed 43; training 180, 187box; use of data 61–2
World Bank 5, 32, 36, 39, 52, 99, 128, 140
World Health Organization *see* WHO
World Summit for Sustainable Development 8
WSP (Water and Sanitation Program) 93

The RiPPLE Programme was funded by the UK Government
from 2006 to 2011.

ukaid
from the British people

From 2006 to 2011, RiPPLE was led by the Overseas Development Institute (ODI), London, UK, together with the following consortium partners:

International Water and Sanitation Centre (IRC), Delft, The Netherlands
Hararghe Catholic Secretariat (HCS), Dire Dawa, Ethiopia
WaterAid Ethiopia, Addis Ababa, Ethiopia
College of Development Studies (CDS) at Addis Ababa University, Ethiopia